T0329160

BARGAINING IN THE SHADOW OF THE MARKET

Selected Papers on Bilateral and Multilateral Bargaining

Kalyan Chatterjee

The Pennsylvania State University, USA

World Scientific

NEW JERSEY · LONDON · SINGAPORE · BEIJING · SHANGHAI · HONG KONG · TAIPEI · CHENNAI

Published by

World Scientific Publishing Co. Pte. Ltd.

5 Toh Tuck Link, Singapore 596224

USA office: 27 Warren Street, Suite 401-402, Hackensack, NJ 07601

UK office: 57 Shelton Street, Covent Garden, London WC2H 9HE

Library of Congress Cataloging-in-Publication Data
Chatterjee, Kalyan.
   Bargaining in the shadow of the market : selected papers on bilateral and multilateral bargaining / by
Kalyan Chatterjee.
    p. cm.
   ISBN-13: 978-9814447560 (hardcover : alk. paper)
   ISBN-10: 9814447560 (hardcover : alk. paper)
   1. Negotiation. I. Title.
   BF637.N4C454 2013
   302.3--dc23

                                         2012046040

British Library Cataloguing-in-Publication Data
A catalogue record for this book is available from the British Library.

In-house Editor: Monica Lesmana

Typeset by Stallion Press
Email: enquiries@stallionpress.com

Printed in Singapore.

BARGAINING IN THE SHADOW OF THE MARKET

Selected Papers on
Bilateral and Multilateral Bargaining

To my wife, Kumkum, and my late parents, Sudhir Ranjan
and Kamala Chatterjee

Preface

The most usual question academics ask each other if they meet after some extended period is "What have you been working on?" Attempting to answer this question for two academic audiences at the University of Birmingham and Rutgers during my sabbatical year of 2007–2008 and for a third at Hitotsubashi University in Tokyo in March 2009, I prepared a talk explaining the motivation and the models in various papers I have been involved in, with several different co-authors, over the years. Whilst I have worked and continue to work in areas other than bargaining, the research in bargaining occupied a major part of my time and seemed to be a reasonably coherent programme of study. An editor at World Scientific, based in New Jersey, saw the slides of this talk posted on the web and contacted me about writing a book based on the lecture. I regarded the prospect of writing a book with some trepidation, especially one that would be an in-depth survey of the extensive literature in a field. My fears were somewhat allayed when the editor assured me that I could stick to the broad outline of my talk, which dealt primarily with my own work with co-authors; expanding and rewriting the talk and including the papers discussed in it would serve as an introduction to some aspects of a vibrant and important research field even though it would not (and does not claim to) be an exhaustive textbook on the topics covered.

I owe a great intellectual debt to the co-authors of papers included in this book: Siddhartha Bandyopadhyay, Gary Bolton, Steve Chiu, Bhaskar Dutta, Ching Chyi Lee, Kathleen McGinn, Debraj Ray, Larry Samuelson, Kunal Sengupta, Tomas Sjöström. My interest in bargaining was first sparked by Howard Raiffa and grew during interactions with him and John W. Pratt. Eric Maskin taught the first course on game theory in the Cambridge, Mass area during my time as a graduate student, which I audited since my oral exams were over, and spent a great deal of time with me after class discussing my work. I am grateful to all of them and to numerous others not mentioned here.

I learnt a great deal also working with Bill Samuelson and Larry Samuelson. I got interested in coalitions as a result of a productive time spent as a visitor at the Indian Statistical Institute, Delhi, working with Bhaskar Dutta, Debraj Ray and Kunal Sengupta.

I would like to thank the economics departments at Rutgers (Tomas Sjöström and Barry Sopher), Birmingham (Siddhartha Bandyopadhyay) and Hitotsubashi (Akira Okada) for giving me the opportunity to try to synthesize the different aspects of the work reported here for lectures to a general economics audience, rather than one solely of theorists. I am grateful to the World Scientific editors, especially Monica Lesmana, for guiding me through this process resulting in the book taking shape.

Finally, I would like to thank my wife, Kumkum, for her unstinting and unquestioning love and support during all these years. When this was first written, she was seriously ill, but alive. Now she too is gone and will see neither the dedication nor this preface.

Acknowledgments

The following publishers are acknowledged for granting permission to re-use my previously published articles:

Elsevier for the following articles:

Kalyan Chatterjee and Ching Chyi Lee, "Bargaining and search with incomplete information about outside options", *Games and Economic Behavior*, 22, 1998, pp. 203–237.

Kalyan Chatterjee and Bhaskar Dutta, "Rubinstein Auctions: On competition for bargaining partners", *Games and Economic Behavior*, 23, 1998, pp. 119–145.

Institute of Electrical and Electronics Engineers (IEEE) for the following article:

Kalyan Chatterjee. Comparison of Arbitration Procedures: Models with complete and incomplete information. *IEEE Transactions on Systems, Man, and Cybernetics*. Vol. SMC–11, No. 2, February 1981. pp. 101–109. Reproduced with permission of Institute of Electrical and Electronics Engineers in the format of reuse in a book/textbook via Copyright Clearance Center.

Nowpublishers for the following article:

Siddhartha Bandyopadhyay, Kalyan Chatterjee and Tomas Sjöström (2011) "Pre-electoral Coalitions and Post-election Bargaining", *Quarterly Journal of Political Science*: Vol. 6: No 1, pp 1–53. http://dx.doi.org/10.1561/100.00010043.

Oxford University Press for the following articles:

Kalyan Chatterjee, Bhaskar Dutta, Debraj Ray and Kunal Sengupta. "A noncooperative theory of coalitional bargaining". *The Review of Economic Studies*, Vol 60, No. 2 (Apr., 2003), pp. 463–477.

Kalyan Chatterjee. "Incentive compatibility in bargaining under uncertainty", *The Quarterly Journal of Economics*, Vol 97, No. 4 (Nov., 1982), pp. 717–726.

The Institute for Operations Research and the Management Sciences (INFORMS) for the following articles:

Reprinted by permission, Gary E. Bolton, Kalyan Chatterjee, Kathleen L. McGinn. How communication links influence coalition bargaining: A laboratory investigation.

Management Science, Vol. 49, No. 5, May 2003, pp. 583–598. Copyright 2003, The Institute for Operations Research and the Management Sciences, 7240 Parkway Drive, Suite 300, Hanover, MD 21076 USA.

Reprinted by permission, Kalyan Chatterjee and Larry Samuelson. Bargaining under two-sided incomplete information: The unrestricted offers case. *Operations Research*, Vol. 36, No. 4, July-August 1988, pp. 605–618. Copyright 1988, The Institute for Operations Research and the Management Sciences, 7240 Parkway Drive, Suite 300, Hanover, MD 21076 USA.

Contents

Preface   vii

Acknowledgments   ix

1. Bilateral and Multilateral Bargaining: An Introduction   1

2. Bargaining and Search with Incomplete Information
   about Outside Options   15

3. Rubinstein Auctions: On Competition for Bargaining Partners   51

4. Bargaining, Competition and Efficient Investment   79

5. A Noncooperative Theory of Coalitional Bargaining   97

6. How Communication Links Influence Coalition Bargaining:
   A Laboratory Investigation   113

7. Pre-electoral Coalitions and Post-election Bargaining   129

8. Comparison of Arbitration Procedures: Models with Complete
   and Incomplete Information   183

9. Incentive Compatibility in Bargaining Under Uncertainty   193

10. Bargaining under Two-Sided Incomplete Information:
    The Unrestricted Offers Case   203

Bilateral and Multilateral Bargaining: An Introduction

Kalyan Chatterjee

1 Motivation for Studying this Field

Most modern markets consist of firms. This is often true both on the production and "consumption" sides. Firms buy from each other, contract with each other, form joint ventures with each other and often share solutions to technological problems with each other (see von Hippel [24]). These transactions are often bilateral, sometimes (as in the case of the sharing arrangement) multilateral in nature. Moreover, even when the transactions are bilateral, they take place in the presence of competition from alternatives on both the buying and the selling sides of the market. A firm is able to get favourable terms if it is "competitive" in the sense of being better than the potential competitors with respect to price or other aspects of the service or good being provided.

An example of negotiations proceeding simultaneously among various players occurs in the industrial gas industry in the 1980s. As on-site production and separation of gases through differential rates of adsorption in a material became the preferred mode of production for many applications, chemical companies (who had the adsorption technology) and industrial gas producers teamed up in joint ventures [1]. The final outcome was British Oxygen forming a match with Dow, Air Liquide with DuPont and Air Products, acquiring the capability by buying Permea from Monsanto. In such a setting, which is not uncommon in the world of industry, there are a small number of players on each side of the market; the process of forming alliances between buyers and sellers of technology takes place by what one could call "competition for bargaining partners". Similarly, it was reported ([19]) that Toyota had rejected a deal with Ford before signing on to the joint venture with GM (whose termination was announced in June 2009).

We thus have bilateral arrangements between buyers and sellers with the expectation that the resulting alliances will compete against each other in the final goods market (for example, from users of industrial gases), an individual buyer seeking out one among two competing sellers, as well as multilateral arrangements among several players.

The basic research questions that arise from market phenomena of this nature are (i) what determines the eventual coalition or coalition structure that forms in equilibrium and (ii) how does competition among potential players on one side of the market affect the payoff of players on the other side in an eventual coalition. These questions motivate the research discussed in the rest of the book.

2 Models

In order to study the theory of bilateral bargaining in the "shadow of the market", so to speak, we need to consider several different kinds of model. The first set of models looks specifically at bilateral bargaining with, sometimes, a given "outside option" representing the effects of competition from other potential partners. We start in Section 1.3, by discussing the classical (and classic) bargaining model of Nash [28], followed by the influential work of Rubinstein ([36]) and the elucidation of the role of outside options in this model in [6]. In Section 1.4, we discuss two papers in which I was involved ([14], [15]), which study bilateral bargaining under incomplete information, and an extension of one of these papers to include outside options in ([22]).

Section 1.5 discusses two papers in which the outside option is not given but has to be obtained through search and one in which it is determined by investment in an incomplete contracts setting . The first paper [13], and its companion [25], models search as a random draw from a probability distribution of offers and considers the question of how the equilibrium is affected by common knowledge of the outcome of search. The second paper [11] starts out with the players in the market all present and their characteristics known, and begins to model competition for bargaining partners (much in line with the motivating story about industrial gas producers and chemical companies).

Section 1.6 considers multilateral bargaining proper, in that meetings are not limited to two players, and seeks to characterise what coalitions will form and how the ones that form share the surplus [12]. We also discuss an experiment that is suggestive about the kind of theory that is appropriate for this problem [7].

The literature mentioned in this chapter focuses primarily on the selected papers in which I was involved, which are reproduced in this volume. This does not aspire to be an exhaustive survey of the various topics. However, the papers themselves contain many more references to related work and texts like [31], [26], [34] and survey articles like [2] also provide more extensive discussions of these topics.

3 Bilateral Bargaining with Complete Information

Modern analysis of the problem of bilateral bargaining began with the two papers of Nash [28], [29]. We discuss this work briefly here, though it is very different in its general approach from the rest of the papers mentioned in this chapter. Nash considered bargaining between two players in a fairly general setup in which the nature of the issue or issues being negotiated were left unspecified. Whatever the issues at stake, a resolution would lead to payoffs for the two players. Measuring these payoffs in von Neumann-Morgenstern expected utilities, the bargaining problem could be represented as the set of feasible utilities for the players. Since these utilities are unique only up to choice of origin and scale, it followed that different choices of these should not affect anything substantive in the description of the bargaining process (with one caveat to follow). Nash also introduced a specific payoff vector, the "status quo" point. Payoffs from the bargaining are essentially measured with this status quo payoff vector as the origin. The status quo is supposed to represent the conflict point or the "best alternative to a negotiated agreement" ([32]). There is (in Nash) no detailed discussion of the right interpretation of the status quo point; later on we shall mention the attempt of Binmore, Rubinstein and Wolinsky to relate this concept to that of "outside option". In his second paper, Nash did attempt to throw some more light on this notion, making it the result of a non-cooperative equilibrium in a particular game.

Given the assumptions on the nature of the set of feasible utilities and the status quo point, Nash proposed some axioms on the "solution" of the bargaining game. It is not very clear what these axioms mean. Raiffa [33] seems to consider them as principles of a fair award, for instance, by an arbitrator. Roth [35] and, probably, Nash himself interpret these axioms to be descriptive of the properties of a class of non-cooperative games. However, this class is not specified beyond the fact that games in it have equilibrium outcomes that correspond to the Nash solution, that is, the solution obtained from Nash's axioms, so it is difficult to take a bargaining game and determine whether it belongs to this class or not. Nash proposed a game, whose Nash equilibria would include his "solution", and a clever refinement argument that would give only his solution. This is known as the Nash demand game. Each of the two players simultaneously and independently writes down a utility demand. If the pair of utilities demanded lies in the feasible set of utilities, there is an agreement and each player obtains his or her demand. If it lies outside the feasible set, there is disagreement and each player gets his status quo payoff. This game is a simple and often realistic model of bargaining, except for the limitation to a single round of simultaneous demands and the possibility of an inefficient agreement if both sides ask for too little — One would anticipate, in the latter instance, that they would renegotiate to an efficient solution on the Pareto frontier. However, it has a continuum of Nash equilibria, so a standard non-cooperative analysis of the game does not yield very much in the way of predictive power. The game has been extended to multiple stages with discounting in [16], which uses strictly extensive form perfect equilibrium as the solution concept, but shows that even this strong refinement notion is unable to reduce the multiplicity. Using a somewhat different notion of perfectness in the one-stage game, Binmore [5] is able to replicate Nash's refinement result.

After Nash, attention in the bargaining literature focused not on the bargaining game, which Nash himself held to be somehow fundamental, but on the axioms and their variants. An elegant discussion of the resulting work is available (up to 1979) in [35]. Rubinstein [36] regenerated interest in the non-cooperative aspects of

bargaining with his path-breaking 1982 work. His main innovation was to introduce
a degree of commitment to every offer, which is missing in any multistage version
of the Nash demand game. Players in his game move sequentially in making offers;
the first proposer is committed to her offer until it is rejected or accepted by the
other player. The other responding player can make a counter-offer after rejecting
the offer he has been given. Thus there is only a single proposal "on the table" at
any given time. Also, in this game, time is supposed to elapse between rejection of
an offer and the making of a counter-offer, so that the responding player in the first
stage has some power to reduce the total benefits by $(1 - \delta)$ where δ is the discount
factor between 0 and 1. When δ is close to 1, the power to reduce the total payoff
is relatively small, but it turns out to still be sufficient. Finally, each player has a
continuum of offers at his or her disposal (as in the demand game or its multistage
variants). Given these three basic features and the use of the subgame perfectness
solution concept, Rubinstein shows that there is a unique solution to the bargaining
problem. Given some properties of preferences, Binmore [4] is able to show that the
utilities obtained in the unique subgame perfect Rubinstein equilibrium converge to
the axiomatic Nash bargaining solution as $\delta \rightarrow 1$, thus providing a striking example
of a bargaining procedure, different from the demand game, which belongs to the
"class" of games "described" by Nash's axioms.

We now turn to the introduction of outside options in Rubinstein's bargaining
game by Binmore, Rubinstein and Wolinsky [6]. The outside option is a utility
payoff a player can get outside the current negotiation being modelled. The authors
distinguish between two models of outside options. In one, a player can choose to
break off the current negotiations, rather than to make a counter-offer, when rejecting a proposal from the other side; she can also choose to take the outside option
rather than to negotiate at all. In this case, the value of the outside option does not
play the role of the disagreement or status quo payoff in Nash's theory; a player's
equilibrium payoff is not strictly increasing in her outside option if it is not large
enough. There is an absolute zero in the utility scale, because of discounting, and
this is the utility of bargaining forever. However, in their second model, Binmore
et al. describe a context where the values of the outside options do act as status
quo payoffs in Nash's sense and the players' utilities are the usual von Neumann-
Morgenstern utilities for risky prospects. In this model, players cannot choose to
break off negotiations; termination of bargaining is an exogenous event with probability $1 - \delta$. If this event occurs, players are forced to take their outside options. In
both instances, the outside options are commonly known and fixed; they are always
available and they do not change. These characteristics, whilst entirely appropriate
for studying the role outside options play in clarifying the relationship between Nash
and Rubinstein bargaining, appear very strong if the focus is on the outside options
themselves and how they give an indication of "bargaining strength". One could
conceive of outside options that need to be generated, either by search or by investment in skills that make an individual bargainer more able to exploit alternatives.
We shall discuss these in more detail in a later section.

4 Bilateral Bargaining with Incomplete Information

Parallel with the development of the Rubinstein model of bargaining, there was a stream of work on bilateral bargaining when each player has proprietary information not available to the other. An early paper was [14]. This paper could be considered an incomplete information version of the Nash demand game. There are two players, a buyer and a seller. Each has a reservation price that is private information; the buyer will not pay more than his reservation price and the seller will not accept less than hers. The probability distribution of reservation prices is common knowledge. The players simultaneously write down bids; if the buyer's is at least as high as the seller's, there is an agreement to trade at a price at a predetermined combination of the bids. If not, there is no agreement and the seller and the buyer both get nothing. There is a multiplicity of equilibria in this but some interest attaches to the equilibria in strictly increasing strategies (except possibly at the boundaries where the probability of agreement is 0, where the strategies need only be weakly increasing). These equilibria are given by a pair of linked differential equations. The paper [14] gives these differential equations and solves them for the special case where the reservation prices are independent draws from a uniform distribution on [0, 1]. Linear strategies are shown to be in equilibrium. It turns out that the possibility exists that there are gains from trade but they are not realised because of aggressive bidding in equilibrium by the players. Such inefficiency is characteristic of the incomplete information condition when the existence of gains from trade is not common knowledge. This latter fact was shown in a restricted class of mechanisms by [9], reproduced in this book, and in a much wider class of mechanisms (though not a superset) in the seminal paper of Myerson and Satterthwaite [27]. Myerson and Satterthwaite also show that the linear equilibrium referred to above maximises the *ex ante* gains from trade.

The simplicity of the [14] paper has rendered it a useful model in applications (see, for example, Chwe [17], Hall and Taylor [23], and Chapter 6 of Fudenberg and Tirole [21]; even though its simplicity also made it quite hard to get published!) It has also found use as a model of double auctions of which multi-player versions serve as good approximations to competitive markets. We shall not discuss this latter avenue of research further here, since the focus in this book is on bargaining.

The problem with the [14] paper as a model of bargaining is that it assumes that players can make and sustain the commitment to stop bargaining at the end of one round of offers. It has been pointed out in many papers that such commitments are often hard to justify. The obvious response from theorists is to consider multiple rounds with sequential proposals *á la* Rubinstein. We focus on the two papers by Chatterjee and L. Samuelson (the second one of which is reproduced in this book). Gantner [22] has recently studied the effect of outside options in one of these models, so this serves as a convenient link with the previous section. For other works, see [2].

The model in the first paper by Chatterjee and L. Samuelson is a concession game. There are two possible types of seller, with reservation price $\bar{s}$ (hard) or $\underline{s}$ (soft) and two types of buyer with reservation prices $\underline{b}$ (hard) and $\bar{b}$ (soft) with the probability of the soft seller (buyer) being denoted by π_s (π_b). The reservation values satisfy $\underline{s} \leq \underline{b} < \bar{s} \leq \bar{b}$. Thus a hard seller and a hard buyer cannot reach a mutually beneficial agreement, whilst a soft type can have a beneficial agreement with either type on the opposite side. One could think of the hard types as "behavioural" types, who consistently stick to the same offer and the soft types as "rational" players, though that is not the interpretation given in the original paper. The only offers permitted in the game are the middle two values, $\underline{b}$, $\bar{s}$ with $\underline{b}$ being the fixed demand of the hard buyer and $\bar{s}$ the corresponding offer of the hard seller. The game in the original paper has the seller start the game by making one of the fixed offers. If she chooses $\underline{b}$ (thus revealing she is the soft type), the game ends. Otherwise the buyer gets to accept or reject. Acceptance of $\bar{s}$ by the buyer ends the game. One could interpret rejection as an offer of $\underline{b}$. In this case, the game moves to the next period and any payoff obtained is discounted by a common discount factor δ. (One can similarly interpret rejection of the buyer's tough offer as the seller herself making a tough offer.) There is a unique Nash equilibrium in this game, which could be either in pure strategies, with immediate concession by one of the players if his or her relative position is sufficiently weak, or in randomised strategies in which each soft type of player randomises between revealing and not revealing. If the action chosen by the random draw is not to reveal, the probability that the player is soft goes down (since the hard type is playing the non-revealing action with probability 1). In finite time, the probability of being soft will hit a threshold. Whoever hits his or her threshold first will then reveal with probability one (if a soft type). Thus the intuitive notion of bargaining strength is captured by the relative values of the threshold probabilities. It is foreseeable who will concede first with probability one, though it is certainly a positive probability event that the other player's randomisation will lead him to reveal first.

5 Search for Outside Options

We will focus here on the papers by Lee [25] and Chatterjee and Lee [13]. However, we first discuss a paper by Gantner [22] that embeds outside options in the first Chatterjee-L. Samuelson paper above. Gantner takes over the basic concession game framework of that paper, including the restriction of possible bargaining offers to be $\{\underline{b}, \bar{s}\}$ and studies how the availability of an outside option to the buyer might change the equilibrium. She discusses outside options obtained from search. The search is modelled as a single-person decision problem for the buyer, who, however, can obtain price offers that are not restricted to $\{\underline{b}, \bar{s}\}$. The buyer receives an offer from the seller and can then choose to accept the offer (by making the same offer himself) or to opt out and search, whilst the seller waits. The search is modelled as a stochastic price offer. The buyer can choose to accept the offer obtained through search, reject it and return to the seller or to continue searching.

With the characteristics of the search distributions being commonly known to both buyer and seller, the search basically acts as a deterrent for the seller to be less aggressive than in the game without the search, if the search option appears to be "good". Otherwise, the buyer leaves. Thus the "rent" to the seller from her private information goes down. Gantner also considers another model, in which the seller does not know the distribution of search options and can only match offers the buyer brings back. Here the buyer has to search to create an outside option where none exists and might return to the bargaining. However, it is not clear why the lack of knowledge about the distribution of search outcomes by the seller is not addressed in a standard Bayesian manner, with the buyer having better information than the seller, rather than in the framework of complete seller ignorance.

The earlier papers by Lee and Chatterjee and Lee (reproduced here) actually address a similar question, about the possibility of both bargaining and search occurring with positive probability in the same equilibrium, but these papers do not study this question in the setting of the incomplete information bargaining game used by her. The basic setup of these models is as follows: Suppose two players are negotiating on a pie of size 1. Seller S makes a price offer p_1; buyer B can accept, reject and ask for another offer, or reject and search at a cost c. Search is modelled non-strategically as taking a draw from a distribution of offers on $[0, 1]$, x_1 being the realised offer. These two papers differ on whether x_1 is observable or not. One would expect a verifiable offer would be more valuable — in fact in one (admittedly surprising) case I know of, a dean of a college asked for the actual offer letter from the other university before coming up with the details of the counter-offer. (The individual concerned took the alternative offer).

A key feature of our model is that offer p_1 is still available. This is akin to the letter with your coming year's salary on it. This recall assumption sets these models aside from other similar ones in the literature and is crucial in obtaining the results. After getting the outside offer, the buyer can now accept this offer, accept p_1 or ask for another offer from the same seller, p_2. Then search again x_2 and choose one offer among p_2, x_1, x_2.

The equilibrium of the model where the search outcome is not observable is as follows — it is unique with a plausible refinement and has pure behavioural strategies on the equilibrium path, and possibly randomised off it:

(1) For $c \geq c^*$, S offers a price that is accepted with probability 1 by the buyer. No search takes place.
(2) For $c < c^*$, the seller offers a price that induces search, buyer searches, if x_1 is low enough buyer accepts it, if it is very high buyer accepts p_1, if in the middle, buyer comes back and asks for another price.
(3) If x_1 is observable, the equilibrium outcome is similar, though with a different cutoff level for the cost of search, $c^{**} > c^*$. So the seller induces search for more values of cost with non-observability of the outside offer.

Interestingly, though perhaps not unexpectedly, the buyer is better off with the outside offer not being observable. This is not because he can claim that it is

higher than it actually is — even if this had been allowed in the game, those with
low outside offers could claim high ones. However, the seller's inability to match an
offer increases the risk of losing the buyer and getting nothing, thus the buyer makes
lower offers on average so that the outside offer is less likely to be more attractive.

The previous model treats alternative offers obtained through search as exogenously given. It makes more sense to try to model a market directly, in which buyers
or sellers choose whom to bargain with on the other side of the market. Such a model
is in [11] (reproduced in this book). We now describe it briefly.

The basic setup is as follows: There is complete information and two buyers
and two sellers in the market. Each seller has one good to sell. The game begins by
sellers simultaneously making offers with each buyer permitted to accept at most one
offer. The buyers too move simultaneously. If at least one offer has been rejected, the
rejectors now make offers simultaneously and the sellers accept or reject, and so on.
Sellers are homogeneous, with value 0. Buyers are heterogeneous with $v_1 > v_2 > 0$.

Three bargaining institutions are considered, namely: (i) public offers; (ii) private
offers; and (iii) targeted offers. In public, untargeted offers, players on each side of
the market (in successive periods) announce prices at which they are willing to sell
or buy respectively. Players on the other side indicate acceptance of at most one
offer. If more than one buyer accepts a single seller's offer, one of them is randomly
chosen to get the good. There is discounting across periods. Public, targeted offers
(such as for example in takeover negotiations) are different in that the price offer
is public but made to a particular player on the other side of the market. Private
offers are also targeted to specific individuals but only the proposer and the recip-
ient actually see the offer. The institution turns out to matter crucially. This is
sometimes seen as a weakness, but should actually be considered a strength of an
extensive form modelling, in that nuances of institutional differences are explicitly observed. However, in most cases, as $\delta \rightarrow 1$, there is a single price in the two
markets, sometimes obtained with one period delay, except in the case of targeted
offers.

With public offers announced without specifically mentioning any particular
agent on the other side, the model gives a pure strategy equilibrium with a uniform
price across buyer-seller pairs, if sellers make offers. If buyers make offers, the situation changes and now there is delay of one period (because there are two pairs of
agents). The price essentially goes to $\frac{v_2}{2}$ as $\delta \rightarrow 1$.

With public, targeted offers, all pure strategy equilibria are shown to involve
delay. If sellers offer first, agreement with the more desirable buyer is delayed for one
period. If buyers offer first, the less desirable buyer waits a period. The important
feature here is that even as $\delta \rightarrow 1$, there are two prices, $\frac{v_2}{2}$ and $\frac{v_1}{2}$. The construction
of this equilibrium rests crucially on a seller making an acceptance-rejection decision
contingent on the offer received by the *other* seller, which seems somewhat counterintuitive.

With private offers, clearly this kind of contingency is not possible. Unfortunately, this also results in the absence of any pure strategy subgame perfect
equilibria. There is a randomised strategy equilibrium in which both sellers offer

with positive probability to the high value buyer and one seller offers with positive probability to both buyers. When buyers make offers, prices essentially converge to $\frac{v_2}{2}$, the lower of the two bilateral bargaining prices. When sellers make offers, however, this is not the case. Prices are such that sellers' *expected* payoffs are $\frac{v_2}{2}$. However, the support of the (equilibrium) mixed strategy does not collapse to a point even for "high" δ.

6 Choice of Investment and Outside Options in Bargaining

One interesting feature of investment in acquiring skills or, in general, human capital, is that the nature of such skills could affect the breadth of the individual's appeal in the job market. The more generalist he or she is, the more his or her options, though in several cases, a specialist might be more productive if the equilibrium assigns him to the right task or partner. A model along these lines is in Chatterjee and Chiu [10](reproduced in this book). There are two buyers and one seller. Each invests in his asset, but the seller's investment is more important for productivity, in a sense made clear in the paper. Contracts on investment cannot be written. Two bargaining formats are examined. In the first, the buyers compete for the seller, as in the Chatterjee-Dutta paper referred to previously. The seller maximises competition by being a generalist and picks up all the gains from trade. This extensive form leads to choice of the optimal level of investment by the seller, given the *type* of investment, but also entails, in equilibrium, the seller choosing a sub-optimal type of investment. The situation is reversed if we adopt a sequential offers extensive form, to be discussed in detail in the next section; there is an underinvestment but of the right type. Which extensive form is better depends on the parameter values, but the trade-off is clear and is intimately related to outside options. Assignment of property rights also plays different roles in the two extensive forms, as expected from the literature on strategic bargaining and property rights, discussed in more detail in the paper.

7 Non-Cooperative Coalitional Bargaining

We now consider a specific game, reminiscent of Rubinstein's alternating offers and Selten's [38] "proposal making model", to answer two questions about coalition formation in the general context of multilateral bargaining. These are: which coalitions will form in equilibrium and how will they divide up the payoff of a coalition among their members. In some sense, this model generalises the discussion of outside options by also allowing "coalitional outside options", in which the payoff depends on being part of an alternative coalition. However, the sequential offers nature of the model precludes much discussion of "competition for bargaining partners" as we shall see later.

We consider first a model with transferable utility and no externalities, with each coalition S obtaining a payoff of $v(S)$ if it forms; there is a fixed protocol in which players move, making and responding to proposals.

A proposal by i consists of a coalition S of which i is a member and a division of $v(S)$ among the members of S. After Player i makes a proposal, members of S respond in the order given by the protocol. If Player j rejects the proposal it is now no longer valid, Player j has the initiative and makes a proposal in the following period. Future payoffs are discounted with a common discount factor δ. This model is different from Rubinstein because of multiple players and different from Selten in two respects, namely (i) the game does not end after a single coalition forms and (ii) there is discounting, so a payoff of 0 has a specific meaning (as in Rubinstein), the utility of bargaining forever without agreement. This model is studied in [12] (reproduced in this book).

Two kinds of results are obtained.

The stationary, subgame perfect equilibria include examples of inefficiency-inefficient delay (though bounded above), inefficient coalition formation for *all* orders of proposers and inefficient coalition formation for *some* order of proposers. Inefficient coalition formation for some order of proposers depends on the returns to additional size of coalitions not being sufficiently large. However, even for games with a high degree of increasing returns to coalition size (namely convex games), there may be some orders of proposal making in which a small sub-coalition forms, even though the grand coalition would have been efficient. It is true, though, that for strictly convex games, there is some order of proposers that gives an efficient coalition, for sufficeintly high values of δ. These negative results essentially survive unless the grand coalition has the highest per capita payoff of all coalitions (a condition called "domination by the grand coalition").

The main positive result, for strictly convex games and sufficiently high δ, is that for the order of proposers for which the efficient coalition (of all N players) forms without delay, the limiting allocation, as $\delta \to 1$ is that point in the core that maximizes the product of players' payoffs. The core and the most egalitarian allocation within it (alternatively a generalized modified Nash bargaining solution) thus appear naturally in the solution of this non-cooperative game. The generalised Nash bargaining solution has also been derived recently by Compte and Jehiel [18], using a condition different from strict convexity of the game. However, they assume Selten's one-coalition property, so their condition is not directly comparable to [12].

We consider an example related to outside options to illustrate some of the negative results.

Example: Consider a three-player game with the following characteristic function: $v(\{1,2\}) = v(\{1,3\}) = 1, v(\{1,2,3\}) = 1 + \epsilon, v(\{S\}) = 0$ otherwise.

For $\epsilon = 0$, there is a unique point in the core with the payoffs to $(1,2,3)$ being $(1,0,0)$. However, the equilibrium of the non-cooperative game gives either $(0.5, 0.5, 0)$ or $(0.5, 0, 0.5)$ as $\delta \to 1$. As ϵ increases, more core points appear, but the (stationary) equilibria of the non-cooperative game give the same payoffs generated by inefficient two-player coalitions $\{1,2\}$ or $\{1,3\}$ until ϵ crosses 0.5, when the grand coalition forms in equilibrium with the payoffs being equal as $\delta \to 1$.

This divergence from the core is somewhat surprising, especially since one might expect players 2 and 3 to compete for Player 1 in this context. This is where the sequential proposals limit Player 1's ability to exploit his or her dominant position. An experimental investigation of a similar game under different communication structures (see Bolton *et al.*, [7] reproduced in this book) finds general support for this kind of behavior though the theory above definitely understates the incidence of formation of the grand coalition relative to the experimental results. Some *ad hoc* explanations are given in Bolton *et al.* [7] to justify this experimental finding, but there is as yet no complete understanding of what feature or features of the extensive form cause this divergence between theory and (one) experiment. The main explanation has to do with the player being left out of the coalition jumping in if he or she sees the other two getting together and accepting a lower payoff then the one he would hold out for according to the model above.

There is also some work on coalition formation with externalities. We refer the reader to the book of Debraj Ray [34] for an excellent exposition of the body of work on this topic, and focus here on an application to political bargaining in Bandyopadhyay *et al.* [3], which is also reproduced in this book. In this paper, we consider elections in parliamentary democracies. Elections in many cases result in bargaining to form a coalition government, in which the surplus is generated by the perquisites of being in government and satisfying the party's constituencies as well as being able to implement the party's policies. Parties that are not in government suffer losses because of policies they oppose being implemented; these losses depend on whether the governing coalition is ideologically close or distant from the particular party that is out of power. The post-election bargaining is strongly influenced by the constitutional rules on which party is selected to be the first proposer, and therefore on the share of seats obtained in the election. The analysis in this part of the paper is similar in spirit to the models already considered in this section, though with externalities for losing parties. We also consider pre-electoral bargaining in which parties form alliances to contest elections, without committing to any particular government or post-election policy. The bargaining here is typically finite horizon, since nominations for parliamentary elections close at a given date after elections are announced, though it could involve many (undetermined) offers and delay does not affect the time at which payoffs are obtained, since this is after the election. These features make the analysis of pre-electoral coalitions somewhat novel. We show in this paper that electoral alliances could result in both proportional representation and first-past-the-post systems, and that a common reason for such alliances is to affect the order of proposers in the post-election bargaining.

8 Conclusion

This chapter tries to put the research reproduced in this book in the context of other work in the area of bilateral and multilateral bargaining. The focus throughout is in trying to understand the bargaining microfoundations of most markets and how institutional differences can be captured in models and make a difference in the predicted results. I have not discussed in detail the paper of mine on arbitration that also appears in this book [8]. This paper focuses also on bargaining between two players but considers the situation where the social costs of disagreement are too high, so that the players voluntarily or the governments concerned decide on a binding arbitration mechanism in the event of a dispute. Unlike earlier papers in this area, which are referred to in the bibliography of my paper, this work is concerned with private information bargaining. One of the models in this is similar to one published a couple of months before this one appeared, by Farber [20], though the two were conceived entirely independently and my paper has other results. Recent papers exploring similar issues are Samuelson [37] and Olszewski [30].

References

[1] Almqvist, Ebbe (2002), *History of Industrial Gases,* Springer, Berlin.

[2] Ausubel, L.M., P. Cramton, and R.J. Deneckere (2002), "Bargaining with Incomplete Information "*Ch. 50 of R. Aumann and S. Hart (eds.) Handbook of Game Theory, Vol 3, Elsevier.*

[3] Bandyopadhyay, Siddhartha, Kalyan Chatterjee and Tomas Sjöström (2011), "Pre-electoral Coalitions and Post-election Bargaining", *Quarterly Journal of Political Science,* 2011, 6, 1–53.

[4] Binmore, K.G. (1987), "Perfect Equilibria in Bargaining Models," in K. Binmore and P. Dasgupta, editors, *The Economics of Bargaining,* Basil Blackwell, Oxford, 1987.

[5] Binmore, K.G., (1981), "Nash Bargaining and Incomplete Information", Discussion paper, University of Cambridge Department of Applied Economics, reprinted in K. Binmore and P. Dasgupta (eds.) *The Economics of Bargaining,* Basil Blackwell, Oxford, 1987.

[6] Binmore, K.G., Ariel Rubinstein, and Ariel Wolinsky (1986), "The Nash Bargaining Solution in Economic Modelling," *RAND Journal of Economics,* 17, 176–188.

[7] Bolton, Gary, Kalyan Chatterjee and Kathleen L. McGinn (2003), "How Communication links coalitional bargaining: A laboratory investigation", *Management Science,* May, 49(5), 583–598.

[8] Chatterjee, Kalyan (1981), "Comparison of arbitration procedures; models with complete and incomplete information", *IEEE Transactions on Systems, Man and Cybernetics,* 11(2), 101–109.

[9] Chatterjee, Kalyan (1982), "Incentive compatibility in bargaining under uncertainty", *Quarterly Journal of Economics,* November, 717–726.

[10] Chatterjee, Kalyan and Y.W. Stephen Chiu (2006), "Bargaining, competition and efficient investment", unpublished manuscript.

[11] Chatterjee, Kalyan and Bhaskar Dutta (1998), "Rubinstein auctions; on competition for bargaining partners", *Games and Economic Behavior,* 23, 119–145, May.

[12] Chatterjee, Kalyan, Bhaskar Dutta, Debraj Ray and Kunal Sengupta, (1993), "A non-cooperative theory of coalitional bargaining", *Review of Economic Studies,* 60(2), 463–477, April.

[13] Chatterjee, Kalyan and Ching-Chyi Lee (1998), "Bargaining and search with incomplete information about outside options", *Games and Economic Behavior,* 22, 203–237, February.

[14] Chatterjee, Kalyan and William Samuelson (1983), "Bargaining under incomplete information", *Operations Research,* 31(5), September-October, 835–851.

[15] Chatterjee, Kalyan and Larry Samuelson (1988), "Bargaining under two-sided incomplete information; the unrestricted offers case", *Operations Research,* 36(4).

[16] Chatterjee, Kalyan and Larry Samuelson (1990),"Perfect equilibria in simultaneous offers bargaining", *International Journal of Game Theory*, 19, 237–267.

[17] Chwe, Michael S-Y (1999), "The reeded edge and the Phillips Curve; money neutrality, common knowledge and subjective beliefs", *Journal of Economic Theory*, 87, 49–71.

[18] Compte, Olivier and Philippe Jehiel (2010), "The Coalitional Nash Bargaining Solution", *Econometrica*, 78, 5, 1593–1623, September 2010.

[19] Copper, John F. (1983), http://www.heritage.org/research/energyandenvironment/bg288.cfm.

[20] Farber, H.S. (1980), "An Analysis of Final-Offer Arbitration", *Journal of Conflict Resolution* December 1980, 24: 683–705.

[21] Fudenberg, Drew and Jean Tirole (1991), *Game Theory*, MIT Press, Cambridge, Mass.

[22] Gantner, Anita (2008), "Bargaining, Search and Outside Options", *Games and Economic Behavior* 62, 417–435.

[23] Hall, Robert E. (2005), "Employment fluctuations with equilibrium wage stickiness", *American Economic Review*, 95(1), 50–65.

[24] von Hippel, Eric (1988), *The Sources of Innovation*, Chapter 6, available online from http://web.mit.edu/evhippel/www/sources.htm.

[25] Lee, Ching Chyi (1994), "Bargaining and Search with Recall: a Two-Period Model with Complete Information", *Operations Research*, 42, 1100–09.

[26] Muthoo, Abhinary (2000), *Bargaining Theory with Applications*, Cambridge, UK: Cambridge University Press.

[27] Myerson, Roger B., Mark A. Satterthwaite (1983). "Efficient Mechanisms for Bilateral Trading", *Journal of Economic Theory* 29(2): 265–281.

[28] Nash, John F. (1950), "The bargaining problem", *Econometrica*, 18: 2 (1950) pp. 155–162.

[29] Nash, John F. (1953), "Two-Person Cooperative Games," *Econometrica*, 21, 128–40.

[30] Olszewski, Wojciech. 2011. "A Welfare Analysis of Arbitration." *American Economic Journal: Microeconomics*, 3(1): 174–213.

[31] Osborne, Martin J. and Ariel Rubinstein (1990), *Bargaining and Markets*, New York: Academic Press.

[32] Raiffa, Howard (1982), *The Art and Science of Negotiation*, Harvard University Press, Cambridge, Mass.

[33] Raiffa, Howard (1953), "Arbitration schemes for generalized two-person games", H.W. Kuhn, A.W. Tucker, and M. Dresher (eds), *Contributions to the theory of games*, 2, Princeton Univ. Press (1953) pp. 361–387.

[34] Ray, Debraj (2007), *A Game Theoretic Perspective on Coalition Formation*, Oxford, UK: Oxford University Press.

[35] Roth, Alvin E. (1979), *Axiomatic Models of Bargaining*, Springer-Verlag.

[36] Rubinstein, Ariel (1982), "Perfect Equilibrium in a Bargaining Model," *Econometrica*, 50, 97–109.

[37] Samuelson, William F. (1991), "Final-Offer Arbitration Under Incomplete Information", *Management Science* 1991 37:1234–1247.

[38] Selten, Reinhard (1981), "A Non-Cooperative Model of Characteristic Function Bargaining," in V. Böhm and H. Nachtkamp (eds.) *Essays in Game Theory and Mathematical Economics*, Mannheim: Bibl. Institut, 131–151. Reprinted in Selten, Reinhard (1989), *Models of Strategic Rationality* Dordrecht, The Netherlands: Kluwer Academic Publishers.

GAMES AND ECONOMIC BEHAVIOR **22**, 203–237 (1998)
ARTICLE NO. GA970586

Bargaining and Search with Incomplete Information about Outside Options

Kalyan Chatterjee*

*Smeal College of Business Administration, Department of Management Science &
Information Systems, The Pennsylvania State University, University Park,
Pennsylvania 16802*

and

Ching Chyi Lee[†]

*Faculty of Business Administration, Department of Decision Sciences and Managerial
Economics, The Chinese University of Hong Kong, Shatin, New Territories, Hong Kong*

Received September 12, 1993

This paper considers a model of bargaining in which the seller makes offers and
the buyer can search (at a cost) for an outside option; the outside option cannot be
credibly communicated, and the seller's offer is recallable by the buyer for one
period. There are essentially two equilibrium regimes. For sufficiently high search
cost, the game ends immediately; otherwise the search occurs in equilibrium.
Compared to the case where the buyer can communicate his outside option, the
seller is worse off, and the game results in search for a smaller set of values of the
search cost, i.e., less equilibrium delay. *Journal of Economic Literature* Classification Number: C72.   © 1998 Academic Press

1. INTRODUCTION AND THE MODEL

The existing literature on extensive form models of bargaining has
focused on two main strategic factors to explain the outcomes of negotiation, namely, incomplete information and time preference. These two

* Dr. Chatterjee acknowledges hospitality from the University of Bonn, and University
College, London, and wishes to thank Avner Shaked and Ken Binmore, respectively, for
bringing this about. We both wish to thank Ken Binmore, Larry Kranich, Abhinay Muthoo,
Motty Perry, two anonymous referees and the Editor, and seminar audiences at the many
places this paper was presented, for useful comments.
† Dr. Lee wishes to thank the Department of MSIS, Penn. State University, for hospitality
during the time a previous version of this paper was written in Summer 1994. Financial
support from the Chinese University of Hong Kong for this research is gratefully acknowledged.

203

categories (which, by the way, have a nonempty intersection) have addressed the issues of disagreement and delay in negotiations between rational players and have sought to draw connections with axiomatically derived solution concepts, like the Nash solution. Examples of the incomplete information work include Chatterjee and Samuelson (1983), Satterthwaite and Williams (1989), Gresik (1991) and the related mechanism design literature of Myerson and Satterthwaite (1983), and others. The time preference aspect is addressed in Rubinstein (1982) and a long line of work proceeding from his paper. Applications tend to use the Rubinstein result, because of its simplicity and its connection with Nash bargaining, even though time preference may not be the real determinant of the allocation of gains from trade in any particular application. This latter point has been made especially forcefully by Hugo Sonnenschein in his presidential address to the Econometric Society in 1989.

A third strategic factor that is generally recognized to be important is a bargainer's ability to generate attractive outside options. (See Raiffa (1982) for a description of the role of "Best Alternatives to the Negotiated Agreement" (BATNA).) This paper and Lee (1994) focus on the analysis of this factor; other recent papers that have done so are Chikte and Deshmukh (1987), Wolinsky (1987), Bester (1988), and Muthoo (1995). There are also related papers that explore the role of fixed, exogenous outside options (Shaked and Sutton, 1984; Binmore, et al., 1989; Shaked, 1987; for example), outside options endogenized through a coalitional bargaining process (Binmore, 1985; Chatterjee and Dutta, 1995, or switching partners (Fudenberg et al., 1987).

The papers by Chikte and Deshmukh (1987), Wolinsky (1987), Bester (1988), and Muthoo (1995) consider bargainers searching for alternatives if the current offer is rejected. Offers cannot be recalled if they are not accepted immediately; as a consequence, search never takes place in equilibrium, since the offerer takes into account a player's search ability in making him or her an offer that cannot be refused.

This finding appears to run counter to the coexistence of search and bargaining in many markets. For example, in the market for senior and middle-level academics, search appears to be a frequently used device to increase the salary or other benefits obtainable from a person's home institution. For those buyers who dislike haggling with car dealers, to give another example, an alternative to the offer/counter-offer process would be to come up with a price quote from either a national pricing service or another dealer.

Both examples above involve recalling offers made by the current negotiating partner as well as outside offers. The scenario at the car dealer's could proceed in the following way, as could the scenarios of other bilateral buyer-seller negotiations. The buyer visits a dealer who first explains the range of cars available at great length. (The delay caused by this is unmodeled in this paper.) The dealer then writes down a price quote for a particular car; the buyer can then make a counter-offer and bargain in the usual Rubinstein (1982) format or leave the dealership to seek an alternative offer. The alternative offer could be from a pricing service (offered, for example, by a well-known credit card company) or a price from another dealer. Either of these has a cost associated with generating it. With the new offer in hand, the buyer can go back and ask for a new offer from the seller. There is some chance that the specific car will have been sold, so that the offer is not recallable. In fact, there is some nonstationarity in the process whereby offers are generated; a second search for a price quote for say, Car X, is more likely to come up blank than the first one, because no more Cars X are available. There is some pressure therefore to conclude agreement. Sometimes there could be a deadline in time, or, say, just two Car X dealers in the vicinity. Search beyond these two then involves a substantially increased cost of travelling to another town, as inhabitants of rural areas will testify.

The car dealer story motivates both the model of this paper and that of Lee (1994). These papers have (i) recall, (ii) one-sided offers, and (iii) a finite (two-period) horizon. Where this paper differs from Lee's companion paper is in the informational assumption about the search outcome—Lee assumes it to be known to both the buyer and the original seller, while this paper considers the case where this is the buyer's private information. Private information about search outcomes could arise if the other dealer refuses to write down a quote. (In the academic market example, this is more common as many schools will not go through the cost of processing an offer without an informal acceptance.) Even if an alternative salary offer is made (in the academic market), this may not reflect the entire offer "package," which would contain subjective elements related to the individual's evaluation of the work and living environment. In the example of the car dealer, the two cars may differ in one or more features, perhaps in color, and an individual's valuation of this difference may be private information. In addition, it is rare that a dealer will put down a written offer in a verifiable format.

The Model

These examples serve to motivate our choice of model. The model is as follows. There are two players, B, the buyer, and S, the seller, with commonly known reservation prices for the item owned by S. These reservation prices are $b = 1$ for the buyer and $s = 0$ for the seller, respectively, so the presence of gains from trade is common knowledge. For convenience, B and S will sometimes be referred to as "he" and "she" in this discussion. S makes an offer p_1 to B in period 1. If B accepts, the payoffs are p_1 to S and $1 - p_1$ to B. B can reject and either search (while still holding on to the offer p_1) or ask for another offer directly without search. (It is immaterial in equilibrium whether the search activity is observed by the seller, since the seller can figure out the equilibrium strategy of the buyer.) However, we assume that search can be observed, even though its outcome cannot.

If B searches at cost c, an offer x_1 is drawn from a commonly known absolutely continuous distribution $F(\cdot)$ with density function $f(\cdot)$. $F(\cdot)$ is assumed to have support on $[0, 1]$. B can now accept x_1, accept p_1, or ask for another offer. In the event of another offer being solicited by B, the offer p_1 lapses and an offer p_2 is made. Either p_2 or x_1 can now be accepted, although there is a small cost of recall—a small probability that x_1 has disappeared. If neither is accepted, B can search again. An independent draw generates x_2 (again at a cost c). Now the buyer has to accept one of p_2, x_1, x_2 or get a payoff of 0. If the buyer at any time accepts x_i, his payoff is $1 - x_i$ – the search cost, while S gets a payoff of 0. On the other hand, if the buyer at any time accepts p_i during period i, his payoff is $1 - p_i$ – the search costs (if any), while S gets p_i. A schematic representation of the extensive form being used is given in Fig. 1.

Depending on the search cost c, this two-period bargaining and search model has two types of equilibrium—one leads to immediate resolution of bargaining, the other involves both bargaining and search. If c is greater than a cutoff c^*, a search deterring offer p^* will be made by S and be accepted by B immediately. For c below c^*, S will offer a price greater than p^*, which will be rejected by the buyer initially and trigger the subsequent search and bargaining behavior.

The intuition behind the delay is as follows. The seller offers a price that is currently unacceptable but may be acceptable later if the outside option is not sufficiently attractive and the continuation expected payoff is decreasing over time. This is better for the seller than offering the maximum acceptable price (search-deterring offer) when c is very low.

A further point of interest in our model is the comparison of complete and incomplete information settings. First we note that, in the absence of any means of verifying outside options, there is no credible way of communicating the buyer's outside option to the seller, so the incomplete information setting is nonvacuous. (The difference between models where cheap talk plays a role (Farrell and Gibbons, 1989, for example), and our model here is that, although the seller wants the buyer to buy from her rather than to go outside, the buyer does not share this interest and just wants the best price, whether it is from an outside option or from the current seller.)

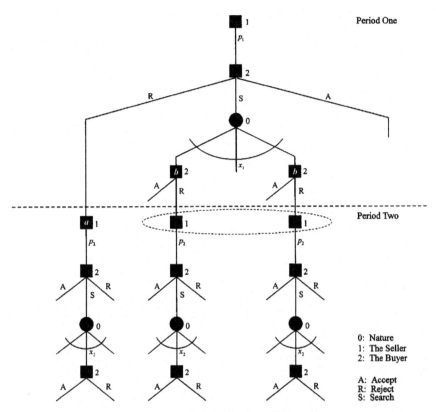

FIG. 1. The extensive form of the search bargaining game.

The complete information model is worked out by Lee (1994). Under complete information as well, there are two types of equilibria, depending on the search cost c. Above a cutoff c^{**}, the price p^* is offered and accepted. Below it, there is search and delay. Interestingly, $c^{**} > c^*$, so that the bargaining ends without delay for a greater range of values of c under incomplete information than under complete information. Furthermore, the seller, on average, is worse off under incomplete information than under complete information. We work out an example in which the buyer is also better off in expected value terms, but we have not been able to prove this in general.

The comparative results might be considered to be somewhat counterintuitive. Why should the buyer be made better off by not being able to communicate his outside option to the seller? The intuition is, however, as follows. In the complete information case, the seller always has the chance to match the buyer's outside option. Here, because of the private information, the seller cannot match the outside option, or the revealed outside option, without creating adverse incentives for the buyer. The chance of "leakage" (a buyer taking his outside option, this being the worst outcome for the seller) makes the seller offer lower prices on average in the low search cost case. (The price p^*, the search deterring price, is the same for both the complete and incomplete information cases.)

The rest of the paper is organized as follow. The next section sets out the equilibria of the two-period model. Section 3 provides discussions on the comparison between complete and incomplete information, along with some comments on extensions to infinite horizon settings. Section 4 concludes. Details of some of the longer proofs are relegated to the appendices.

2. THE ANALYSIS

Before we start the analysis, let us first describe the equilibria of the game. We restrict ourselves to equilibria involving pure strategies on the equilibrium path. The perfect Bayesian equilibria we obtain have the following characteristics on the outcome path. For a cost c above a cutoff c^*, S will offer a price p^* that is immediately accepted by the buyer. The price p^* is the largest price the seller can charge without inducing search by the buyer. The sufficient condition that ensures the above equilibrium is given later in Proposition 6A. Under some conditions on the distribution of outside offers and on what we shall argue to be implausible beliefs, however, a price lower than p^* could be charged (which also leads to immediate resolution of bargaining). This is discussed in Proposition 6B. Thus, for sufficiently high values of c, there could be nonuniqueness of equilibria; however, only one will survive the plausibility restriction.

For a cost c below the cutoff value c^*, we get a unique equilibrium in which the seller finds it optimal to "take her chances" by charging a price p_1 higher than p^* in the beginning. The buyer initially rejects and undertakes a search. If the search option x_1 is in the highest of three regions, the buyer accepts p_1. If it is in the intermediate region, the buyer asks for a second offer. In the lowest region, the buyer accepts the outside option and the game ends. If the second period is ever reached, the seller then charges another price p_2 higher than p^*, (interestingly, $p_2 > p_1$ in this equilibrium), and the buyer searches again. In this equilibrium, search and bargaining both occur, because of both "recall" and the fact that the game ends after two rounds. In addition, only pure strategies are used along the equilibrium path, although mixed strategies are needed off the equilibrium path to support the equilibrium. A formal description of this equilibrium is given in Proposition 7.

BARGAINING AND SEARCH WITH INCOMPLETE INFORMATION   209

For clarity of the presentation, Σ is used to denote the complete game described in the previous section. In addition, Σ^a and Σ^b. are defined as follows:

Σ^a: The subgame starting at node a in Fig. 1 (i.e., the subgame starting at the beginning of the second period, given that the buyer does not search in the first period). Σ^a is essentially a single-period version of the full model.

Σ^b: The subgame starting at node b in Fig. 1 (i.e., the subgame starting at the end of the first period after the buyer has searched and has obtained an x_1).

We start our analysis by examining each player's behavior in Σ^a and Σ^b. Since the buyer's optimal behavior in the second period is the same in both Σ^a and Σ^b, we shall consider this case first.

2.1. *The Buyer's Behavior in the Second Period*

Let $s = \min(x_1, p_2)$. Notice that, for Σ^a, since the buyer does not search in period 1, $s = p_2$. For Σ^b, since the outside option x_1 is not observable by the seller, s is the buyer's private information.

Let $\Phi(s)$ denote the buyer's expected payoff if he searches in the second period. Then, since recall of previous offers is allowed,

$$\Phi(s) = -c + \int_0^s (1-x)f(x)\,dx + \int_s^1 (1-s)f(x)\,dx. \qquad (1)$$

It is easy to see that $\Phi(s)$ is decreasing in s and takes its maximum value, $1 - c$, at $s = 0$, and its minimum value, $1 - c - \int_0^1 xf(x)\,dx$, at $s = 1$. We assume that $\Phi(s)$ is nonnegative throughout, so c is not "too large."

Now let p^* be the price that makes the buyer indifferent between searching and not searching. Then p^* must satisfy the following condition:

$$1 - p^* = -c + \int_0^{p^*} (1-x)f(x)\,dx + \int_{p^*}^1 (1-p^*)f(x)\,dx \equiv \Phi(p^*).$$

$$(2)$$

Since (2) can be rewritten as $c = \int_0^{p^*} F(x)\,dx$, there exists one and only one p^* satisfying (2). We now have the following lemma, which will be quite useful for our later analysis.

LEMMA 1. *Let $s \in [0, 1]$. Then $1 - s > \Phi(s), \forall s < p^*$ and $1 - s < \Phi(s), \forall s > p^*$.*

Proof. We can write $(1 - s) - \Phi(s) = c - \int_0^s F(x)\,dx$. Note that the expression is equal to c (> 0) at $s = 0$ and, since $\int_0^s F(x)\,dx$ is increasing in s, that it remains positive until $c - \int_0^s F(x)\,dx = 0$, at $s = p^*$. The same reasoning tells us that it then becomes negative for $s > p^*$.   ∎

Lemma 1 simply says that a search will take place in the second period if and only if $\min(x_1, p_2) > p^*$. Hence p^* is the cutoff price for a search to occur in the second period. We shall show later that p^* is also the cutoff price for a search to take place in the first period. In fact, Lemma 1 implies that, in any period within a multiperiod model, no price (including both the seller's and outside offers) larger than p^* will be accepted immediately. This is because, if the price is higher than p^*, the buyer can search at least once more, which yields an expected payoff higher than immediate acceptance of the price.

Given the buyer's optimal strategy in the second period, we can characterize the equilibrium of Σ^a and Σ^b. We shall start with Σ^a, the simpler case.

2.2. The Equilibrium of Σ^a

Since the buyer does not search in the first period, and since p_1 lapses when p_2 is offered, the seller's strategy in the second period does not depend on the history of Σ in the first period. The equilibrium then depends only on c. In Σ^a, the seller can offer p^*, knowing it will be accepted (by Lemma 1), and get a payoff of p^* (she certainly will not offer anything less than p^*); or she can offer a price p_2 greater than p^*, knowing the buyer will search (again, by Lemma 1) and expect to get $p_2[1 - F(p_2)]$ (since p_2 will only be accepted when $x_2 > p_2$).

Let $\bar{p}$ maximize $p_2[1 - F(p_2)]$. (An assumption is made below to guarantee that $\bar{p}$ is unique.) Then the seller's optimal choice of p_2 in Σ^a will be either p^* or $\bar{p}$, depending on which of p^* and $\bar{p}[1 - F(\bar{p})]$ is larger. In equilibrium, the seller offers p^* if $p^* \geq \bar{p}[1 - F(\bar{p})]$, and otherwise offers $\bar{p}$. In either case, the buyer's expected payoff for going to the second period directly without search in the first period is at most $1 - p^*$. We illustrate this in Fig. 2.

For technical reasons and to facilitate comparison of this model to the corresponding complete information case, the following assumption is made.

Assumption. $p[1 - F(p)]$ is strictly quasi-concave. This assumption guarantees that $p[1 - F(p)]$ is single-peaked and contains no flat portion in the interior of $[0, 1]$. Among the frequently used probability functions,

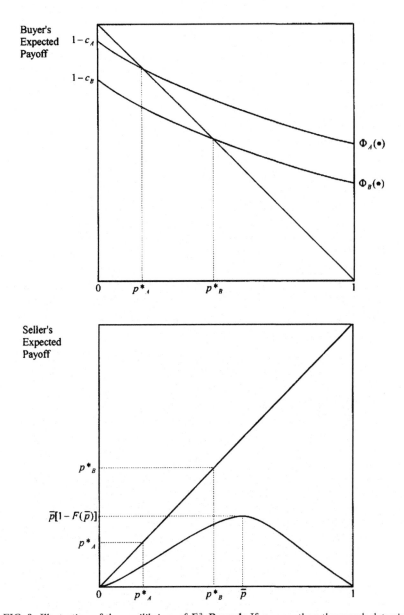

FIG. 2. Illustration of the equilibrium of Σ^a. **Remark:** If $c = c_A$, then the search-deterring price is p_A^*. In this case, since $p_A^* < \bar{p}[1 - F(\bar{p})]$, the seller will offer $\bar{p}$, and the buyer will search. If $c = c_B$, then the search-deterring price is p_B^*. In this case, since $p_B^* > \bar{p}[1 - F(\bar{p})]$, the seller will offer p_B^*, and the buyer will accept immediately.

for example, the family of power functions, $F(x) = x^m$ $(m > 0)$, satisfies this assumption.

2.3. *The Equilibrium of* Σ^b

The equilibrium of Σ^b is more complicated because it depends not only on c but also on p_1 and x_1. In addition, since x_1, is not observable by the seller, the seller must have a belief about x_1, and that belief, in equilibrium, must be consistent with the buyer's behavior in the first period.

There are two types of equilibria for Σ^b—one involves search (in the second period), the other does not. By Lemma 1, for the no-search equilibrium to arise, a price less than or equal to p^* must be offered by the seller in the second period, whereas for the equilibrium to involve a search, a price greater than p^* must be offered. Since p^* is increasing in c, the type of equilibrium that will arise then depends on c. The larger c is, the larger p^* is, and hence the higher is the seller's incentive for offering a search deterring offer. For c below a cutoff value, the seller will strictly prefer offering a price higher than p^* to offering a price below, and vice versa for c above the cutoff. In addition, the equilibrium also depends on p_1. If $p_1 \le p_2 \le p^*$, then, intuitively, the second period should not be reached since no better offer from the seller is expected in the second period. The following proposition considers this case.

PROPOSITION 1.   *If* $p_1 \le p^*$, *the equilibrium strategies of the buyer and the seller in* Σ^b *are as follows: The buyer accepts* p_1 *at the end of the first period if* $x_1 > p_1$, *accepts* x_1 *if* $x_1 \le p_1$. *The seller offers* $p_2 \ge p_1$ *in the second period. The seller's beliefs are such that* p_2 *is optimal given these beliefs and that* $p_2 \le p^*$. *The buyer accepts immediately the minimum of* x_1 *and* p_2 *in the second period.*

Remark.   This proposition is placed here for completeness. It covers two cases, which will be dealt with in detail in Propositions 6A and 6B. In one case, argued to be the only plausible case, $p_1 = p_2 = p^*$, and the seller's belief (given the buyer returns for the second period) is that x_1 is greater than p^*. Another case could potentially arise in which a price p° strictly less than p^* is optimal in the second period. Note that from Lemma 1, a second-period search cannot occur in either of these equilibria.

The remaining discussion on the equilibria of Σ^b will only consider the case when $p_1 < p^*$. Proposition 2 describes a no-search equilibrium of Σ^b under such a condition.

PROPOSITION 2.   *Suppose* $p_1 > p^*$. *Suppose also* $p^* \ge$ *max* $p_2[1 - F(p_2)]^2/[1 - F(p^*)]$. *Then, the unique equilibrium of* Σ^b *is the following: The seller, believing* $x_1 \in (p^*, 1]$ *whenever the second period is reached,* *offers p^* in the second period. The buyer accepts x_1 at the end of the first period if $x_1 \leq p^*$; otherwise he goes to the second period to accept p^* if $x_1 > p^*$. If the offer p_2 is strictly greater than p^*, as is x_1, the buyer searches and chooses the minimum of x_1, x_2, and p_2.*

Proof. Given that $x_1 \in (p^*, 1]$, the seller has the following two options in the second period. First, she can offer a price $p_2 = p^*$ to get a sure payoff of p^*. Second, she can offer a p_2 greater than p^*, knowing that the buyer will search, and expect to get

$$p_2 \cdot \text{Prob}[\, p_2 < x_1 \text{ and } p_2 < x_2 \,|\, x_1 \in (p^*, 1]\,]$$

$$= \frac{1 - F(p_2)}{1 - F(p^*)} p_2 [1 - F(p_2)]$$

Clearly, if $p^* \geq \max p_2[1 - F(p_2)]^2/[1 - F(p^*)]$, the seller will offer p^* in the second period.

Since the conjectured second period offer is p^*, the buyer will go to the second period to accept p^* if and only if $x_1 > p^*$. ∎

We now turn our attention to the case where there is second-period search by the buyer. Before constructing the equilibrium, let us make the following four observations:

First, if the equilibrium of Σ^b is to involve search, then, as we have argued before (by Lemma 1), the seller's second period offer must be greater than p^*.

Second, for the seller to offer $p_2 > p^*$, it must be the case that the expected payoff from offering such a price is better than the payoff from offering the search-deterring price p^*. This implies that the search cost c is sufficiently low for the seller to take her chances.

Third, since the conjectured second period offer is greater than p^* (according to the first observation), the buyer will go to the second period if and only if $x_1 > p^*$. Therefore, the seller's belief about x_1 must have p^* as the lower bound for second-period search to occur. (Note that a buyer with $x_1 > p^*$ will consider it optimal to go through a second search rather than accepting the outside offer. This follows from the definition of p^*.)

Hence, assuming that the seller believes $x_1 \in (p^*, b]$ ($p^* < b \leq 1$) whenever the second period is reached (we will verify later that the seller's belief must indeed be in this form in equilibrium), then for the seller to offer $p_2 > p^*$, it must be the case that

$$\max p_2 \cdot \text{Prob}[\, p_2 \text{ will be accepted} \,|\, x_1 \in (p^*, b]\,] > p^*$$

$$\Leftrightarrow \max p_2 \cdot \text{Prob}[\, p_2 < x_1 \text{ and } p_2 < x_2 \,|\, x_1 \in (p^*, b]\,] > p^*$$

$$\Leftrightarrow \max \frac{F(b) - F(p_2)}{F(b) - F(p^*)} p_2 [1 - F(p_2)] > p^*.$$

For simplicity, define $\Psi(p_2, b)$ as follows:

$$\Psi(p_2, b) \equiv \frac{F(b) - F(p_2)}{F(b) - F(p^*)} p_2[1 - F(p_2)]. \tag{3}$$

Furthermore, let $\hat{p}(b)$ be the maximizer and $\Psi(\hat{p}(b), b)$ be the maximum value of $\Psi(p_2, b)$ for a given b.

Notice that $\hat{p}(b)$ and $\Psi(\hat{p}(b), b)$ have the desirable properties described in the following lemma, which we will need in the sequel.

LEMMA 2. *Define* $\mathbf{B} \approx \{b | \Psi(\hat{p}(b), b) \geq p^*, \text{ and } p^* < b \leq 1\}$. *Then, for all* $b \in \mathbf{B}$, *both* $\hat{p}(b)$ *and* $\Psi(\hat{p}(b), b)$ *are continuous and monotonically increasing in* b.

Proof. See Appendix A. ■

Finally, our last observation concerns the relationship between p_1 and p_2 in any equilibrium that involves second-period search. Suppose $1 - p_1 > \Phi(p_2)$. Then $\Phi^{-1}(1 - p_1) < p_2$, and all buyers with $x_1 > \Phi^{-1}(1 - p_1)$ will accept p_1. Then the seller's belief about x_1 must have support on $[p^*, \Phi^{-1}(1 - p_1)]$, of which p_2 is not a member. Therefore p_2 is clearly not optimal. (See Fig. 3 for the above discussion.) Therefore, $1 - p_1 \leq \Phi(p_2)$ in any equilibrium with second-period search.

If $1 - p_1 < \Phi(p_2)$, then p_1 will be rejected by all buyers with $x_1 \in (p^*, 1]$, and, therefore, p_2 must maximize $\Psi(p_2, 1)$, where

$$\Psi(p_2, 1) = \frac{1 - F(p_2)}{1 - F(p^*)} p_2[1 - F(p_2)]. \tag{4}$$

To further simplify the notation, let $\hat{p} = \hat{p}(1)$, i.e., $\hat{p}$ maximizes (4). Then, when $1 - p_1 < \Phi(\hat{p})$, the condition $\hat{p}[1 - F(\hat{p})]^2/[1 - F(p^*)] > p^*$ (which is exactly the opposite condition of Proposition 2) gives rise to the unique equilibrium of Σ^b, in which the seller offers $p_2 = \hat{p}$ and the buyer searches. Proposition 3 below describes such an equilibrium.

PROPOSITION 3. *Suppose* $1 - p_1 < \Phi(\hat{p})$ *and* $\hat{p}[1 - F(\hat{p})]^2/[1 - F(p^*)] > p^*$. *Then the following is the unique equilibrium of* Σ^b: *The buyer accepts* x_1 *at the end of the first period if* $x_1 \leq p^*$; *otherwise, expecting* $p_2 = \hat{p}$, *he goes to the second period to search if* $x_1 \in (p^*, 1]$. *The seller, believing* $x_1 \in (p^*, 1]$ *whenever the second period is reached, offers* $p_2 = \hat{p}$.

Proof. See Fig. 4 and previous discussions. ■

The preceding proposition delineates the equilibrium of Σ^b under the conditions that (i) $1 - p_1 < \Phi(\hat{p})$ and (ii) $\hat{p}[1 - F(\hat{p})]^2/[1 - F(p^*)] > p^*$.

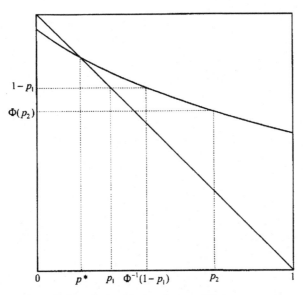

FIG. 3. Illustration of the relationship between equilibrium p_1 and p_2. **Remark:** If $1 - p^* > 1 - p_1 > \Phi(p_2)$, then the buyer will accept x_1 if $x_1 \leq p^*$, will go to the second period to search if $x_1 \in (p^*, \Phi^{-1}(1 - p_1)]$, and will accept p_1 if $x_1 > \Phi^{-1}(1 - p_1)$. Hence, if the second period is reached the seller should believe that $x_1 \in (p^*, \Phi^{-1}(1 - p_1)]$, of which p_2 is not a member.

The first condition implies $p_1 > p^*$. Notice that if the second condition is violated, we will then have the equilibrium described in Proposition 2.

If the first condition is violated, i.e., if $1 - p_1 > \Phi(\hat{p})$ (which, combined with the condition that $p_1 > p^*$, implies $p^* < p_1 < 1 - \Phi(\hat{p})$), then $\hat{p}$ can no longer be the optimal second-period offer. In fact, based on our previous observation that $1 - p_1 \leq \Phi(p_2)$ and the property of $\Phi(\cdot)$, a price smaller than $\hat{p}$ will be offered by the seller in the second period. In this case, there are then two possible second-period offer scenarios. We can have $1 - p_1 = \Phi(p_2)$, in which case those buyer types indifferent between accepting p_1 and waiting for p_2 must accept or reject in the "right" proportion. That is, assuming that the support of the seller's belief is an interval rather than a union of disjoint intervals, a buyer with $x_1 \in (b, 1]$ must accept p_1 and $x_1 \in (p^*, b]$ must wait for p_2. The quantity b, which is unique if it exists, is determined jointly by the following three conditions: (i) p_2 maximizes $\Psi(p_2, b)$, i.e., $p_2 = \hat{p}(b)$; (ii) $\Psi(\hat{p}(b), b) \geq p^*$; and (iii) $1 - p_1 = \Phi(\hat{p}(b))$. The first two conditions guarantee that $\hat{p}(b)$ is the seller's optimal choice of offer, given $x_1 \in (p^*, b]$ (in fact, if (ii) holds for equality, the seller is indifferent between offering p^* and $\hat{p}(b)$, in

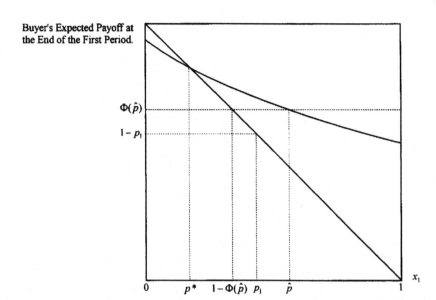

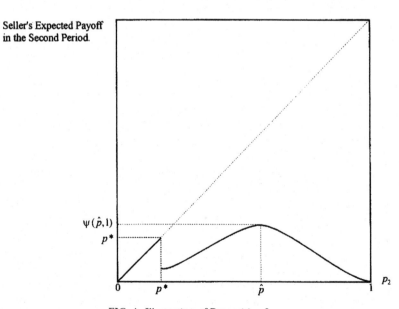

FIG. 4. Illustration of Proposition 3.

which case the equilibrium requires her to offer $\hat{p}(b)$) and the last condition ensures that the buyer will return to the second period if and only if $x_1 \in (p^*, b]$. More specifically, if $x_1 \in (p^*, p_1)$, the buyer strictly prefers going to the second period. If $x_1 \geq p_1$, however, the buyer is indifferent between accepting p_1 at the end of the first period and going to the second period to search again. In this case, the equilibrium requires the buyer to accept p_1 if $x_1 > b$ and to go to the second period if $x_1 < b$. Refer to Fig. 5 for the above discussion.

Notice that, because of the continuity and monotonicity of $\hat{p}(b)$ and $\Psi(\hat{p}(b), b)$ described in Lemma 2, we can always find the "b" that satisfies conditions (i) and (iii) above. However, the $\Psi(\hat{p}(b), b)$ so constructed may not satisfy (ii), and then the $\hat{p}(b)$ so determined cannot be the equilibrium second offer. This situation can arise when p_1 is relatively small, so that $\hat{p}(b)$ and $\Psi(\hat{p}(b), b)$ are also small. In this case, an alternative scenario becomes operative, in which the cutoff b is chosen differently so that (i) $\hat{p}(b)$ maximizes $\Psi(p_2, b)$; (ii) $\Psi(\hat{p}(b), b) = p^*$; and (iii) $1 - p_1 = \alpha(1 - p^*) + (1 - \alpha)\Phi(\hat{p}(b))$. (The existence of $\hat{p}(b)$ and $\Psi(\hat{p}(b), b)$ that satisfy all of the above conditions is guaranteed by Lemma 2.) The first two conditions indicate that the seller is indifferent between offering p^* and $\hat{p}(b)$ in the second period when $x_1 \in (p^*, b]$. The third condition then says that she will offer each price with probabilities α and $1 - \alpha$, respectively. The last condition, for a similar reason explained in the previous scenario, also ensures that the buyer will return to the second period if and only if $x_1 \in (p^*, b]$. Refer to Fig. 6 for the above discussion.

To summarize, when $P_1 \in (p^*, 1 - \Phi(\hat{p})]$, depending on the magnitude of p_1, one of the above two scenarios will arise. The first scenario arises for larger values of p_1 and the second for smaller values of p_1. There exists a cutoff that separates these two situations.

Based on the above discussions, we now have the following propositions. Proofs are omitted.

PROPOSITION 4. *Suppose $p_1 > p^*$ and $1 - p_1 \geq \Phi(\hat{p})$. If there exists a unique quantity "b" such that (i) $\hat{p}(b)$ maximizes $\Psi(p_2, b)$; (ii) $\Psi(\hat{p}(b), b) \geq p^*$; and (iii) $1 - p_1 = \Phi(\hat{p}(b))$; then the unique equilibrium of Σ^b is as follows: The buyer accepts x_1 if $x_1 \leq p^*$, accepts p_1 if $x_1 > b$, and goes to the second period and searches if $p^* < x_1 \leq b$. The seller believing that $x_1 \in (p^*, b]$ whenever the second period is reached, offers $p_2 = \hat{p}(b)$.*

PROPOSITION 5. *Suppose $p_1 > p^*$ and $1 - p_1 \geq \Phi(\hat{p})$. If the unique quantity "b" that satisfies the three conditions of Proposition 4 does not exist, then there must exist a b' such that (i) $p(b')$ maximizes $\Psi(p_2, b')$; (ii) $\Psi(\hat{p}(b'), b') = p^*$; and (iii) $1 - p_1 = \alpha(1 - p^*) + (1 - \alpha)\Phi(\hat{p}(b'))$. The unique equilibrium of Σ^b is as follows: The buyer accepts x_1 if $x_1 \leq p^*$,*

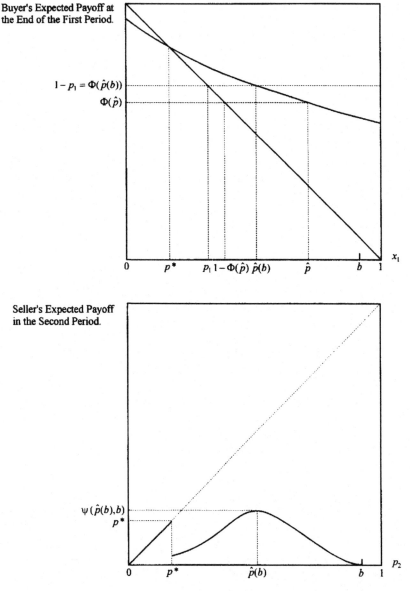

FIG. 5. Illustration of Proposition 4. **Remark:** Given $p_1 < -\Phi(\hat{p})$, the seller offers $\hat{p}(b)$ if $\psi(\hat{p}(b), b) > p^*$, such that $1 - p_1 = \Phi(\hat{p}(b))$. Hence, if $x_1 \leq p^*$, the buyer will accept x_1 at the end of the first period; if $x_1 \in (p^*, \hat{p}(b))$, the buyer strictly prefers returning to the second period; and if $x_1 \geq \hat{p}(b)$, the buyer is indifferent between accepting p_1 at the end of the first period and returning to the second period. In equilibrium, if the buyer is indifferent between accepting p_1 and returning to the second period, he will accept p_1 if his outside offer, x_1, is greater than b

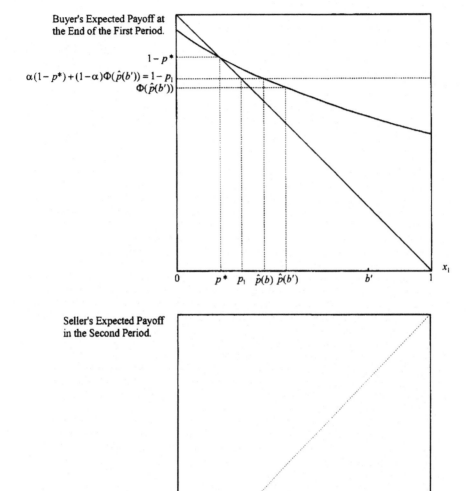

FIG. 6. Illustration of Proposition 5. **Remark:** Here, given p_1, it is not optimal for the seller to offer $\hat{p}(b)$ such that $1 - p_1 = \Phi(\hat{p}(b))$, since $p^* > \psi(\hat{p}(b), b)$. Hence the seller randomizes in the second period between offering p^* and $\hat{p}(b')$ (with probabilities α and $1 - \alpha$, respectively) such that $\alpha(1 - p^*) + (1 - \alpha)\Phi(\hat{p}(b')) = 1 - p_1$. If $x_1 \leq p^*$, the buyer will accept x_1 immediately; if $x_1 \in (p^*, \Phi^{-1}(1 - p_1))$, the buyer strictly prefers returning to the second period; and if $x_1 > \Phi^{-1}(1 - p_1)$, the buyer is indifferent between accepting p_1 and returning to the second period, in which case he accepts p_1 if $x_1 > b'$.

accepts p_1 if $x_1 > b'$, and goes to the second period to search if $p^* < x_1 \le b'$.
The seller, believing that $x_1 \in (p^*, b']$ whenever the second period is reached,
offers $p_2 = \hat{p}(b')$ with probability α and $p_2 = p^*$ with probability $1 - \alpha$.

Propositions 1–5 delineate equilibrium behavior in Σ^b for different
values of c and p_1. It can be checked that these propositions have indeed
considered all possible values of c and p_1. It is explicitly assumed above
that, no matter what the values of c and p_1 are, the buyer finds it optimal
to search in the first period (otherwise, the subgame Σ^b will not be
reached.) In the following analysis of the equilibrium of the full model, we
will examine the consistency of such an assumption. We shall show that
only when $\hat{p}[1 - F(\hat{p})]^2/[1 - F(p^*)] > p^*$ and $p_1 > p^*$ will the first
period involve search. Hence, if these two conditions are met, the scenarios described in Propositions 1 and 2 will appear only off the equilibrium
path.

2.4. *The Equilibrium of* Σ

As in both Σ^a and σ^b, there are two types of equilibria for Σ,
depending on the search cost—one involves search, the other does not. If
a condition similar to that of the optimality of p^* in the second period is
satisfied, then our intuition seems to suggest that the seller should offer
$p_1 = p^*$ also in the first period (since there is no reason for the seller to
delay offering p^*, knowing it will be accepted by the buyer). If this is true,
then, search can only occur off the equilibrium path, and the seller's
second-period offer will be used to support her equilibrium first-period
offer.

We can now check that p^* (possibly along with some other prices less
than p^*) will, in fact, be offered in the first period in equilibrium for
relatively high values of c. We state this formally in the following proposition.

PROPOSITION 6A. *Suppose $p^* \ge \hat{p}[1 - F(\hat{p})]^2/[1 - F(p^*)]$. Then the
following is an equilibrium of Σ. The seller offers $p_1 = p^*$ in the first period.
If the buyer searches in the first period and returns, the seller, believing
$x_1 \in (p^*, 1]$, offers p^* in the second period. If the buyer does not search in the
first period, then the seller either offers $p_2 = p^*$, if $p^* \ge \bar{p}[1 - F(\bar{p})]$, or
offers $p_2 = \bar{p}$, if $\bar{p}[1 - F(\bar{p})] > p^*$. The buyer accepts any offer less than or
equal to p^* when offered. In equilibrium, the game ends immediately, with the
seller getting p^* and the buyer getting $1 - p^*$, respectively.*

Proof. See Appendix B. ■

It should be noted that, if $\bar{p}[1 - F(\bar{p})] < p^*$, we could have an equilibrium in which the seller makes a trivial offer $p_1 > p^*$ in the first period

and the buyer simply rejects it and goes to the second period directly to accept p^* in the second period. Such an equilibrium would yield exactly the same payoffs for both players as in the equilibrium of Proposition 6A.

Thus, given the conditions of Proposition 6A, it is an equilibrium for the seller to offer p^* in both periods, and for the buyer to accept immediately. However, this is not the only possible equilibrium without search. The next proposition demonstrates that, depending on the probability distribution and with an off-equilibrium belief different from that of Proposition 6A, an equilibrium could exist with the price charged by the seller being below p^*.

PROPOSITION 6B. *Let p° be a price less than p^*. Suppose that (i) $p^\circ > p_2[1 - F(p_2)]/[1 - F(p^\circ)]$ for all $p_2 \in (p^\circ, p^*]$; and (ii) $p^\circ > p_2[1 - F(p_2)]^2/[1 - F(p^\circ)]$ for all $p_2 \in (p^*, 1]$. Then the following is an equilibrium of Σ: The seller offers $p_1 = 1 - \Phi(p^\circ)$ in the first period. If the buyer searches in the first period, the seller, believing $x_1 \in (p^\circ, 1]$, offers p° in the second period. If the buyer does not search in the first period, then the seller either offers $p_2 = p^*$, if $p^* \geq \bar{p}[1 - F(\bar{p})]$, or offers $p_2 = \bar{p}$, if $\bar{p}[1 - F(\bar{p})] > p^*$. The buyer in the first period, accepts any offer less than or equal to $1 - \Phi(p^\circ)$, and searches otherwise. If x_1, the outcome of first period search, is less than p°, the buyer accepts x_1. If $x_1 \geq p^\circ$, the buyer asks for a second offer from the seller. In the second period, if $\min(x_1, p_2) > p^*$, the buyer searches; otherwise, he accepts the minimum of x_1 and p_2 without search. In equilibrium, the game ends immediately, with the seller getting $1 - \Phi(p^\circ)$ and the buyer getting $\Phi(p^\circ)$, respectively.*

Proof. See Appendix C. ∎

The intuition behind the above equilibrium is simple. If the seller believes that there is a probability that the buyer may enter the second period with an outside offer less than p^*, then it may be optimal for her to offer a price less than p^*.

In the above equilibrium, if the seller believes that the buyer may enter the second period with $x_1 \in (p^\circ, 1]$, then she will do one of the following three things in the second period. First, she can offer $p_2 = p^\circ$ to get a sure payoff of p° (since $p^\circ < p^*$, it will be accepted immediately.) Second, she can offer a price between p° and p^*, i.e., $p_2 \in (p^\circ, p^*]$. In this case, since $p_2 < p^*$, the offer will be accepted by the buyer immediately if it is less than x_1. Hence the seller's maximum expected payoff for making such an offer will be $\max p_2[1 - F(p_2)]/[1 - F(p^\circ)]$, where $[1 - F(p_2)]/[1 - F(p^\circ)]$ is the conditional probability that $x_1 > p_2$ given $x_1 \in (p^\circ, 1]$. Finally, the seller can offer a price greater than p^*. In such a case, the buyer will search and the offer will not be accepted unless both x_1 and x_2 are greater than p_2. Hence the most the seller can get for taking this last option is max $p_2[1 - F(p_2)]^2/[1 - F(p°)]$, where $[1 - F(p_2)]^2/[1 - F(p°)]$ is the conditional probability that both x_1 and x_2 are greater than p_2, given $x_1 \in (p°, 1]$.

It is certainly possible that, for some forms of $F(\cdot)$ and some values of $p°$, the first option will yield the highest expected payoff for the seller and, hence, a price smaller than p^* will be offered in the second period. If this is the case, then the seller in the first period will also have to offer a price smaller than p^*. The reason why the seller does not have to offer $p_1 = p°$ and offers $p_1 = 1 - \Phi(p°)$ (which is greater than $p°$) instead is that the buyer will have to search in the first period to expect $p°$ in the second period. Hence, to deter search, the seller need only offer a price such than $1 - p_1 = \Phi(p°)$.

Notice that offering p^* in both periods is also an equilibrium strategy for the seller, given the conditions of Proposition 6B, if p^* is the optimal second period offer, given that a buyer with x_1 in $(p^*, 1]$ returns for a second period. This equilibrium is more plausible in the following sense. Suppose the equilibrium first period offer is as in Proposition 6B and the seller deviates and offers p^*. The equilibrium can be sustained by the buyer belief that this deviation is a "mistake" and that the second period offer will be $p°$, in which case the deviating offer will be rejected. However, p^* is the equilibrium offer in an alternative perfect Bayes equilibrium, and the deviation can be interpreted not as a mistake, but as a signal of the seller's intention to offer the same price in the second period. The buyer should believe this signal, because it is the opening choice in a perfect Bayes equilibrium that gives the seller a higher payoff. If the buyer believes this, the seller should certainly deviate; this destroys the candidate (Proposition 6B) equilibrium. Henceforth, we shall disregard the equilibrium outlined in Proposition 6B.

Both types of equilibria described above require high search costs. As the cost of search c becomes smaller, the value of p^* decreases and the seller's expected payoffs in these equilibria also decrease. Eventually, when c becomes small enough, it may no longer be optimal for the seller to resolve bargaining "immediately." In such cases, the seller, as seen in the subgame Σ^b, may then try to take her chances by offering something higher than p^*, knowing that the buyer will search and hoping that her offer will be accepted eventually because of the buyer's unfruitful search. This can happen when the condition for Proposition 6A does not hold, i.e., when $p^* < \hat{p}[1 - F(\hat{p})]^2/[1 - F(p^*)]$. We now consider these cases.

When analyzing the equilibria of Σ^b in the previous section, it is explicitly assumed that, for values of p_1 given in the propositions (1–5), the buyer finds it optimal to search in the first period. We now show that, under the condition $p^* < \hat{p}[1 - F(\hat{p})]^2/[1 - F(p^*)]$, only when $p_1 > p^*$

will the buyer indeed find it optimal to search. If the seller offers anything lower than p^*, the buyer will accept immediately, and hence the second period will never be reached.

Consider first if the seller offers $p_1 > p^*$ in the first period. Then, as implied by Lemma 1, it is never optimal for the buyer to accept immediately. This is because the buyer can only get $1 - p_1$ if he accepts immediately, whereas if he searches he can expect to get at least $\Phi(p_1)$ (because he can always "recall" p_1 at the end of the first period). Since $p_1 > p^*$, $\Phi(p_1) > 1 - p_1$. Clearly, accepting p_1 immediately in such a case is not optimal for the buyer.

If the buyer goes to the second period directly without searching in the first period, then the subgame Σ^a will be reached. According to our previous analysis, since $p^* < \hat{p}[1 - F(\hat{p})]^2/[1 - F(p^*)]$ (which implies $p^* < \bar{p}[1 - F(\bar{p})]$ by simple algebra), the seller will offer $\bar{p}$ in the second period, and the buyer's expected payoff for taking this action will be $\Phi(\bar{p})$. However, if the buyer does search in the first period, the subgame Σ^b will be reached and one of the situations depicted in Propositions 3 through 5 will arise. This implies that the buyer can expect to get at least $\Phi(\hat{p})$, because in this case the second period seller offer will be no larger than $\hat{p}$. Lemma 3 below shows that $\hat{p} < \bar{p}$. Since $\Phi(\cdot)$ is a decreasing function, it is clear that a search is the best response for the buyer in the first period if $p_1 > p^*$.

LEMMA 3. *Let $\bar{p}$ and $\hat{p}$ be the prices that maximize $p[1 - F(p)]$ and $p[1 - F(p)]^2$, respectively. Then $\bar{p} \geq \hat{p}$.*

Proof. See Appendix D. ∎

We next show that if $p^* < \hat{p}[1 - F(\hat{p})]^2/[1 - F(p^*)]$ and the seller offers $p_1 \leq p^*$, then the buyer's best response is to accept immediately.

If $p_1 \leq p^*$, the buyer can either (1) accept immediately to get $1 - p_1$, (2) search, in which case the scenario described in Proposition 1 will arise and no offer better than p_1 is expected from the seller in the second period (but the buyer has paid the search cost), or (3) go to the second period immediately without a search, in which case his expected payoff will be $\Phi(\bar{p})$, as discussed before. Clearly, since $p_1 \leq p^*$ and $1 - p_1 > \Phi(p_1) > \Phi(\bar{p})$, immediate acceptance of p_1 is optimal for the buyer.

Hence, given the condition that $p^* < \hat{p}[1 - F(\hat{p})]^2/[1 - F(p^*)]$, the buyer's equilibrium strategy in the first period will be to immediately accept any seller offer smaller than or equal to p^* and to search otherwise.

Now we have come to the last part of the analysis—the seller's optimal choice of p_1. The preceding analysis suggests that the seller's optimal choice of p_1 must be either p^*, in which case the seller can ensure a payoff of p^*, or a price greater than p^*, in which case, depending on the size of p_1, one of the three cases described in Propositions 3–5 will arise. Let us define the sets of p_1 that induce the different kinds of subgame behavior described in Propositions 3, 4, and 5 as $\mathbf{P_U}$, $\mathbf{P_M}$, and $\mathbf{P_L}$, respectively. Notice that $\mathbf{P_U}$, $\mathbf{P_M}$, and $\mathbf{P_L}$ together exhaust the interval $(p^*, 1]$ and are mutually exclusive (except in the boundary case). Furthermore, the values of p_1 in $\mathbf{P_U}$ are larger than those in $\mathbf{P_M}$, which, in turn, are larger than those in $\mathbf{P_L}$.

Consider first the cases when $p_1 \in \mathbf{P_U}$. That is, consider the case when the seller's first period offer is such that $p_1 > 1 - \Phi(\hat{p})$. In this case, since $p_1 > p^*$, the buyer will search and the situation depicted in Proposition 3 will arise. Since, according to Proposition 3, the buyer will return to the second period if and only if $x_1 > p^*$ and, once the second period is reached, the seller will offer $p_2 = \hat{p}$ to expect a payoff of $\hat{p}[1 - F(\hat{p})]^2 / [1 - F(p^*)]$, the seller's expected payoff for offering such a p_1 will be

$$\int_{p^*}^{1} \frac{1 - F(\hat{p})}{1 - F(p^*)} \hat{p}[1 - F(\hat{p})] f(x_1) \, dx_1 = \hat{p}[1 - F(\hat{p})]^2, \tag{5}$$

which is independent of p_1. In other words, if $p_1 > 1 - \Phi(\hat{p})$, the value of p_1 becomes irrelevant because p_1 will never be accepted. However, notice that if the seller offers $p_1 = 1 - \Phi(\hat{p})$ (which is the largest member in $\mathbf{P_M}$), she will have exactly the same expected payoff (by Proposition 4). We therefore eliminate from consideration offers p_1 in $\mathbf{P_U}$, as these are payoff-equivalent to the largest offer in $\mathbf{P_M}$.

Consider next the cases when $p_1 \in \mathbf{P_L}$, i.e., those values of p_1 that induce the subgame behavior described in Proposition 5. Then, according to Proposition 5, p_1 will be accepted if and only if $x_1 \in (b', 1]$, and the second period will be reached only with probability $F(b') - F(p^*)$, where b' is chosen so that $\Psi(\hat{p}(b'), b') = p^*$. Furthermore, once the second period is reached, the seller will randomize between offering p^* and $\hat{p}(b')$, and hence her expected payoff in the second period will be p^*. As a result, the seller's expected payoff for offering such a p_1 will be

$$p_1[1 - F(b')] + p^*[F(b') - F(p^*)] \tag{6}$$

Since b' in (6) does not depend on p_1 (see Proposition 5), (6) is increasing in p_1 as long as the value of p_1 stays in $\mathbf{P_L}$. Hence, if the seller is to offer a p_1 in this range, she will offer the largest p_1 possible. Notice that, by condition (ii), which determines the value of b' in Proposition 5, given p^*, b' and $\hat{p}(b')$ are uniquely determined. Hence, $\Phi(\hat{p}(b'))$ in (iii) of Proposition 5 must also be fixed. Thus, when the value of p_1 increases, to maintain the equality in (iii) of Proposition 5, the value of α must decrease because $1 - p^* > 1 - p_1 \geq \Phi(\hat{p}(b'))$ (see Fig. 6). Let p'_1 be the largest price in $\mathbf{P_L}$. Then it must be true that $1 - p'_1 = \Phi(\hat{p}(b'))$ and $\alpha = 0$. This then means that p^* will be chosen with probability zero in the second period. Hence, in equilibrium, the seller will never play a mixed strategy, even if she may be indifferent between offering p^* and $\hat{p}(b')$ in the second period. The seller's expected payoff for offering $p_1 = p'_1$, according to (6), is

$$p'_1[1 - F(b')] + p^*[F(b') - F(p^*)].$$

However, it can be checked that the p'_1 so defined is also a member of $\mathbf{P_M}$ (the smallest one). In fact, p'_1 is the price such that condition (ii) of Proposition 4 holds for equality. We therefore can also rule out any $p_1 \in \mathbf{P_L}$ (except p'_1) as the possible equilibrium first-period offer.

Hence the first-period equilibrium offer must be either p^* or a price in $\mathbf{P_M}$. According to Proposition 4, if the seller offers $p_1 \in \mathbf{P_M}$, p_1 will be accepted if and only if $x_1 \in (b, 1]$, and the second period will be reached with probability $F(b) - F(p^*)$. Furthermore, since the seller will offer $\hat{p}(b)$ to expect a payoff of $\Psi(\hat{p}(b), b)$ once the second period is reached, the seller's expected payoff for offering such a p_1, denoted as $S'(p_1)$, will be

$$S'(p_1) = p_1[1 - F(b)] + [F(b) - F(p^*)]\Psi(\hat{p}(b), b), \qquad (7)$$

where b (and hence $\hat{p}(b)$ and $\Psi(\hat{p}(b), b)$) is chosen so that the three conditions of Proposition 4 are satisfied.

Let $\tilde{p}(\in \mathbf{P_M})$ be the maximizer (assumed unique) and $S'(\tilde{p})$ be the maximum value of (7). Then, in equilibrium, the seller will offer either p^* or $\tilde{p}$ in the first period, depending on which of p^* and $S'(\tilde{p})$ is larger.

We now summarize the above discussion in the following proposition.

PROPOSITION 7. *If $p^* < \hat{p}[1 - F(\hat{p})]^2/[1 - F(p^*)]$, then the unique equilibrium of Σ is as follows: If $p^* \geq S'(\tilde{p})$, the equilibrium calls for the seller to offer p^* and the buyer to accept immediately in the first period. If $p^* < S'(\hat{p})$, the seller offers $\tilde{p}(> p^*)$ and the buyer searches in the first period. The subsequent subgame behavior of both players along the equilibrium path is described in Proposition 4.*

The seller's ex ante equilibrium expected payoff, denoted as $S(p)$, is

$$S(p) = \max\{p^*, S'(\tilde{p})\}.$$

The buyer's ex ante equilibrium expected payoff denoted as $B(p)$, *is*

$$B(p) = \begin{cases} 1 - p^* & \text{if } p_1 = p^* \\ B'(\tilde{p}) & \text{if } p_1 = \tilde{p}, \end{cases}$$

where

$$B'(\tilde{p}) = -c + \int_0^{p^*} (1 - x_1) f(x_1) \, dx_1 + \int_{p^*}^{\hat{p}(b)} \Phi(x_1) \, dx_1$$

$$+ \int_{\hat{p}(b)}^1 (1 - \tilde{p}) f(x_1) \, dx_1.$$

Proof. See previous discussion. $B'(\tilde{p})$ can be verified by referring to the buyer's behavior described in Proposition 4. ∎

We make some remarks about the uniqueness of the equilibrium. Clearly, if the game has multiple equilibria, it must be because there are different off-equilibrium beliefs that can support different equilibrium second offers. Hence, when checking the uniqueness of the equilibrium, we can thus concentrate only on the seller's belief about x_1 in the second period, and see whether there are beliefs other than those specified previously that can support different equilibria.

If $p^* > \hat{p}[1 - F(\hat{p})]^2 / [1 - F(p^*)]$, as we pointed out in Propositions 6A and 6B, the equilibrium may not be unique, depending on the form of $F(\cdot)$, although we did argue that the belief used to support the equilibrium in Proposition 6B is not plausible. However, both equilibria call for immediate resolution of bargaining, even if the value of the equilibrium first-period offer is different.

If $p^* < \hat{p}[1 - F(\hat{p})]^2 / [1 - F(p^*)]$, as required by Proposition 7, then, from our previous discussion, we know that there is no belief that can support $p_2 \leq p^*$ in equilibrium, i.e., the equilibrium p_2 must be greater than p^*. Hence, if p^* maximizes $S(p)$, the only possible equilibrium first offer for the seller is p^*, and therefore, the equilibrium is unique. If $\tilde{p}$ maximizes $S(p)$, by assuming that (in the second period) the seller belief about x_1 is an interval, the equilibrium in the subsequent subgame is always unique, regardless of whether the seller follows the equilibrium path (which leads to Proposition 4) or deviates (which leads to Proposition 1, 3, or 5). The equilibrium in this case is also unique.

3. DISCUSSION

The analysis in the preceding section suggests that the incomplete information model has two types of equilibria (namely, "search" and "no search" equilibria) that are similar qualitatively to those of the complete information model of Lee (1994). A comparison between players' *ex ante* expected payoffs in this model and those in Lee's model shows the following two interesting propositions. To help visualize these results, we construct an example in which $F(\cdot)$ is uniform, and the players' *ex ante* equilibrium expected payoffs in this case are plotted in Fig. 7.

PROPOSITION 8.   *The seller's ex ante expected payoff under incomplete information is always less than or equal to her ex ante expected payoff under complete information.*

Although this result is intuitive (the seller is better off to be informed rather than uninformed), its proof is not trivial; it is shown in Appendix E.

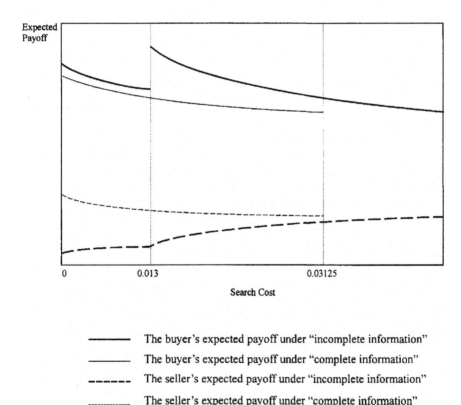

FIG. 7. The seller's and the buyer's *ex ante* equilibrium expected payoffs when $F(\cdot)$ is uniform. When $c > 0.03125$, the expected payoff under incomplete information is the same as that under complete information for both players. This is the case when "no search" equilibrium arises—the seller offers p^* and the buyer accepts. The figure is not drawn to scale.

Hence, not being able to observe the buyer's outside option makes the seller worse off. A parallel conclusion about the buyer's *ex ante* expected payoff under each model cannot be drawn. It is not clear whether the buyer is surely better off in one model and worse off in the other. However, as shown in Fig. 7, when $F(\cdot)$ is uniform, the buyer is better off in the incomplete information case.

The reason a definitive comparison of the buyer payoffs is difficult to make in general is that under complete information, the seller in the second period may more than match the buyer's first-period outside offer. For example, if the buyer comes to the second period with $x_1 \in (p^*, p'']$, where $p'' = \inf\{p \mid p[1 - F(p)] \geq p^*\}$, the seller will offer $p_2 = p^*$, and the buyer will accept. See Lee (1994) and Appendix E for details. This phenomenon cannot occur under incomplete information.

PROPOSITION 9. *The seller will offer the search-deterring price p^* for a larger set of c under incomplete information than under complete information.*

This result follows directly from Proposition 8 and is proved in Appendix E. Readers can also refer to Fig. 7 for this result. In Figure 7, it is shown that the cutoff c that separates "search" equilibrium from "no search" equilibrium in the incomplete information model is 0.013, while the cutoff in the complete information model is 0.03125. Hence, surprisingly, immediate resolution of bargaining is more likely to occur under incomplete information than under complete information. To put it differently, incomplete information bargaining achieves the efficient outcome for a larger domain of values of c.

The intuition behind these results is as follows. Since the buyer is unable to go to the seller with a verifiable outside option and ask for a matching or lower offer, there is more "leakage" from the system than there is with complete information. That is, the buyer accepts a good outside option, since to try to use it to generate a competing seller offer is not credible. The seller compensates for this by generally making lower demands in equilibrium, and by offering the search-deterring price for a greater range of costs than under complete information. The buyer, except in the one case mentioned above, does not lose by not being able to convey outside offers credibly to the seller, because if the seller does match, the buyer is at most only slightly worse off with his outside option than with the seller offer.

A further point of interest for the incomplete information model is that the second-period seller offer is always greater than the first one if the search equilibrium arises. This can be seen from a condition for search equilibrium that $1 - p_1 = \Phi(p_2)$ for equilibrium p_1 and p_2. Since both p_1 and p_2 in this equilibrium must be greater than p^*, this condition ensures that $p_2 \leq p_1$ in equilibrium. This result is intriguing because it is observed quite frequently in real life, but not very often in theoretical models. We often hesitate to buy a product when the price is low and end up having to pay a higher price for the same good.

Our model consists of only two periods. With a finite horizon (beyond two periods), the qualitative features of the above result will remain, although the cutoff values on cost may be different. With an infinite horizon, however, the delay result may or may not hold, depending on the features of the environment. If the process is stationary (i.e., with constant search cost), the infinite horizon stationary equilibrium will not involve search. The price p^* will be offered at every period and accepted in the first period. Thus recall alone is not sufficient to achieve delay in an infinite horizon model; there must be other features of the environment that interact with it to introduce some nonstationarity in the bargaining and search process. For example, we can show that, if the cost of search goes up with the number of searches, then the infinite horizon model may also involve search and delay.[1] Thus recall and some degree of nonstationarity, caused by finiteness or by increasing costs over time, play a major role in determining whether an equilibrium with search exists, as in our model, or one with just bargaining exists, as in the other papers cited in the introduction. Thus the two-period setting used here is not essential to obtain equilibrium search.

4. CONCLUSIONS

In this paper we have analyzed a two-period model of bargaining and search with recall, the result of search being assumed to be the searcher's private information. The equilibria of the model could involve search with

---

[1] Consider the following example, where $c_1 < c_2 = c_3 = \cdots$. Then there will exist a $p_2^* = p_3^* = \cdots$, such that in every stationary equilibrium, the buyer will reject any price higher than p_2^* from time period 2 onward. Therefore, in period 2, the seller will offer p_2^*, where p_2^* satisfies $1 - p_2^* = \Phi_2(p_2^*)$ as before, and the game will end in two periods. We examine the seller's first-period offer. Clearly, since $c_1 < c_2$, the first-period search-deterring offer p_1^* will be less than p_2^*. It may be optimal for the seller to offer something between p_1^* and p_2^* in period 1. This depends on whether the seller's payoff is maximized by offering p_1^*, with a payoff of p_1^* or $p_1[1 - F(p_1)]$, where $p_1 \in (p_1^*, p_2^*)$. Since p_2^* depends on $c_2, \ldots, c_n$, finding such a p_1 for sufficiently large c_2 is the same as determining the cutoff value c_1^*, such that for $c_1 < c_1^*$, the equilibrium with bargaining and search both occurring will hold. (For further elaboration, see Chatterjee and Lee (1996), an earlier version of this paper.) A referee has pointed out, however, that the best way of modeling these features is not to regard the search process as exogenously specified, but as a process by which a new bargaining round starts if a suitable partner is found. This paper should be considered an initial attempt at addressing these issues rather than the last word on the subject; the referee's comment points to the need for further elaboration in future research.

probability 1 when the search cost is sufficiently small, as well as immediate acceptance of a search-deterring price when the search cost is sufficiently large. The outside option, being private information, without any credible means of communicating this information to the other side, carries with it the initially somewhat surprising implication that the searcher is, in general, better off, and the party without information (the seller) is worse off than in the complete information case.

The recall feature could be used in equilibrium, either just before the end of the game, or just before the end of the first period, when the seller's offer is about to lapse. In the latter case, the recall is not of an outside offer, but that of a price to which the offering party is committed for one period.

In the decision-theoretic search literature, recall is potentially used only in the last period of a finite-horizon game, and never in an infinite-horizon setting, given the independence assumptions we have about outside offers. We argue here that the recall assumption, along with some kind of finiteness of the game, is necessary to achieve equilibria with both bargaining and search.

APPENDIX A: PROOF OF LEMMA 2

We first note that if $b \in \mathbf{B}$, then $\hat{p}(b)$ must be in the interior of $(p^*, b]$. This is easy to check.

The continuity of both $\Psi(\hat{p}(b), b)$ and $\hat{p}(b)$ follows from the "maximum theorem" (see Harris, 1987, for example). The monotonicity of $\Psi(\hat{p}(b), b)$ is demonstrated below by showing $\Psi(\hat{p}(b'), b') > \Psi(\hat{p}(b''), b'')$ for all $b' > b''$:

$$
\begin{aligned}
\Psi(\hat{p}(b''), b'') &= \hat{p}(b'')[1 - F(\hat{p}(b''))] \frac{F(b'') - F(\hat{p}(b''))}{F(b'') - F(p^*)} \\
&\leq \hat{p}(b'')[1 - F(\hat{p}(b''))] \frac{F(b'') - F(\hat{p}(b''))}{F(b') - F(p^*)} \\
&< \hat{p}(b')[1 - F(\hat{p}(b'))] \frac{F(b') - F(\hat{p}(b'))}{F(b') - F(p^*)} \\
&= \Psi(\hat{p}(b'), b').
\end{aligned}
$$

The second inequality follows from the fact that $\hat{p}(b')$ is the unique maximizer of (3) for $b = b'$, and the first from the fact that $\hat{p}(b'') > p^*$.

We now prove that $\hat{p}(b)$ is increasing in b. The proof is by contradiction. Suppose $\hat{p}(b)$ is not increasing in b; then there must exist b' and b'' such that $b' > b''$ and $\hat{p}(b') < \hat{p}(b'')$. To simplify our notation, we now let $p' = \hat{p}(b')$ and $p'' = \hat{p}(b'')$. Then

$$\frac{F(b'') - F(p'')}{F(b'') - F(p^*)} p'' [1 - F(p'')] > \frac{F(b'') - F(p')}{F(b'') - F(p^*)} p' [1 - F(p')]$$

$$\Leftrightarrow F(b'')\{p''[1 - F(p'')] - p'[1 - F(p')]\}$$
$$> F(p'')p''[1 - F(p'')]$$
$$- F(p')p'[1 - F(p')] \qquad (i)$$

By the same reasoning, we have

$$F(b')\{p'[1 - F(p')] - p''[1 - F(p'')]\}$$
$$> F(p')p'[1 - F(p')] - F(p'')p''[1 - F(p'')] \qquad (ii)$$

Adding (ii) to (i), we get

$$[F(b'') - F(b')]\{p''[1 - F(p'')] - p'[1 - F(p')]\} > 0$$

Since $F(b'') - F(b') < 0$, we must have

$$p''[1 - F(p'')] < p'[1 - F(p')]. \qquad (iii)$$

Now, let $\bar{p} = \arg\max_{p \in [0, 1]} p[1 - F(p)]$ as before. By the assumption of strict quasi-concavity of $p[1 - F(p)]$, if we can show that both p' and p'' are less than $\bar{p}$, we can conclude that (iii) is a contradiction.

We now show that p' and p'' are both indeed less than $\bar{p}$.

Suppose $p'' > \bar{p}$, and $b'' < 1$, then $F(b'') - F(\bar{p}) > F(b'') - F(p'')$. Hence,

$$\frac{[F(b'') - F(\bar{p})] \bar{p}[1 - F(\bar{p})]}{F(b'') - F(p^*)} > \frac{[F(b'') - F(p'')] p''[1 - F(p'')]}{F(b'') - F(p^*)}.$$

However, this is a contradiction, since p'' is suppose to maximize the right-hand side of the above expression. Hence $p'' < \bar{p}$. A similar argument can be used to show that $p' < \bar{p}$.

By the assumption that $p[1 - F(p)]$ is strictly quasi-concave, $p''[1 - F(p'')] < p'[1 - F(p')] \Leftrightarrow p'' < p'$, which contradicts our original assumption.

Therefore, for $b' > b''$, we must have $p' > p''$, i.e., $\hat{p}(b)$ is increasing in b. ∎

APPENDIX B: PROOF OF PROPOSITION 6A

We start our proof by examining the players' behavior in the second period.

The buyer's strategy in the second period follows directly from Lemma 1.

Consider the case when the subgame Σ^a is reached (i.e., the buyer does not search in the first period). Then, by our discussion in Section 2.2, the seller will offer either p^* or $\bar{p}$, depending on which of p^* and $\bar{p}[1 - F(\bar{p})]$ is larger.

Now consider the case when the subgame Σ^b is reached (i.e., the buyer searches in the first period). Then, by Proposition 2, since $p^* > \hat{p}[1 - F(\hat{p})]^2/[1 - F(p^*)]$, the seller will offer p^* in the second period. The buyer, after observing x_1 at the end of the first period, will accept $\min[x_1, p_1]$ if $\min[x_1, p_1] < p^*$; otherwise, he will go to the second period to accept the seller's offer, p^*.

Now, consider the beginning of the first period. It should be clear that the seller will never offer anything strictly less than p^* (by Lemma 1) if we use the off-equilibrium belief described in Proposition 1.

Suppose the seller offers a price $p_1 > p^*$. In this case, the buyer has the following three options. First, he can accept p_1 immediately to get $1 - p_1$. Second, he can reject p_1 and go directly to the second period to get either $1 - p^*$ or $\Phi(\bar{p})$ (see discussion on Section 2.2). Third, he can search in the first period, leading to an expected payoff of $\Phi(p^*) = 1 - p^*$ (since the second-period offer in this case is expected to be p^*). By the decreasing property of $\Phi(\cdot)$ and the definition of p^*, it is clear that the buyer's first alternative is never optimal.

If $\bar{p}[1 - F(\bar{p})] > p^*$, the buyer's expected payoff for taking the second action is $\Phi(\bar{p})$, which is less than his expected payoff for taking the third action $\Phi(p^*)$, because $\bar{p} > p^*$. Hence, if $\bar{p}[1 - F(\bar{p})] > p^*$, the buyer's best response to $p_1 > p^*$ is to search in the first period, leading to a seller offer of $p_2 = p^*$ and an expected payoff of $p^*[1 - F(p^*)]$ for the seller (because the second period will be reached with only probability $1 - F(p^*)$). The seller is therefore better off offering p^* immediately, thus ensuring a payoff of p^*.

If $\bar{p}[1 - F(\bar{p})] < p^*$, the buyer's payoff for taking the second action is $1 - p^*$, which makes him indifferent between the second and third actions. In this case, if the buyer takes his second action, the seller's payoff is p^*, which is the same as her payoff if she offers p^* directly. However, if the buyer assigns some positive probability for choosing his third action, then the seller will get a payoff that is strictly less then p^*. Again, the seller is better off to offer p^* immediately to ensure a payoff of p^*. ■

APPENDIX C: PROOF OF PROPOSITION 6B

The buyer's strategy in the second period follows directly from Lemma 1. The seller's strategy in the second period if the buyer does not search in the first period follows from our discussion in Section 2.2.

If the subgame Σ^b is reached (i.e., the buyer has searched in the first period), then the seller will choose a price p_2 to maximize his second-period expected payoff $S_2(p_2)$, conditional on the belief that $x_1 \in [p^\circ, 1]$, where

$$
S_2(p_2) = \begin{cases}
p_2 & \text{if } p_2 \in [0, p^\circ] \\[2mm]
\dfrac{p_2[1 - F(p_2)]}{1 - F(p^\circ)} & \text{if } p_2 \in (p^\circ, p^*] \\[4mm]
\dfrac{p_2[1 - F(p_2)]^2}{1 - F(p^\circ)} & \text{if } p_2 \in (p^*, 1].
\end{cases}
$$

By assumptions (i) and (ii) of the Proposition, it is clear that p^* maximizes $S_2(p_2)$ in the second period.

Now, at the end of the first period, after observing x_1, the buyer will certainly accept $\min[x_1, p_1]$ if $\min[x_1, p_1] < p^\circ$; otherwise, the buyer will return to the second period and accept $p_2 = p^*$.

Now, consider the beginning of the first period. If the seller offers $p_1 > 1 - \Phi(p^\circ)$ (note that $p^\circ < 1 - \Phi(p^*)$ since $p^\circ < p^*$), the buyer's payoff is $1 - p_1$ if he accepts, $\Phi(p^*)$ if he searches (since he expects to get p^* in the second period), and either $1 - p^*$ or $\Phi(\bar{p})$ if he goes to the second period directly without search (depending on whether $p_2 = p^*$ or $p_2 = \bar{p}$). Clearly, the buyer in this case should search. As a result, the seller's expected payoff for offering $p_1 > 1 - \Phi(p^*)$ is $p^\circ[1 - F(p^\circ)]$.

If the seller offers $p_1 \leq 1 - \Phi(p^\circ)$, the buyer's payoff is $1 - p_1$ if he accepts, $\Phi(\min(p_1, p^\circ))$ if he searches, and either $1 - p^*$ or $\Phi(\bar{p})$ if he goes to the second period directly without search. Clearly, the buyer's best response is to accept. Given this, the seller can guarantee herself a payoff of $1 - \Phi(p^\circ)$ if she offers $p_1 = 1 - \Phi(p^\circ)$, while her payoff is only $p^\circ[1 - F(p^\circ)]$ if she offers $p_1 > 1 - \Phi(p^\circ)$ (it is straightforward to check that $1 - \Phi(p^\circ) > p^\circ[1 - F(p^\circ)]$). Hence, in equilibrium, the seller offers $p_1 = 1 - \Phi(p^\circ)$, and the buyer accepts immediately. ■

APPENDIX D: PROOF OF LEMMA 3

Suppose $\bar{p} < \hat{p}$. Then $F(\bar{p}) \leq F(\hat{p})$, and hence,

$$
1 - F(\bar{p}) \geq 1 - F(\hat{p}). \tag{i}
$$

We know that

$$\bar{p}[1 - F(\bar{p})] > \hat{p}[1 - F(\hat{p})]. \qquad (ii)$$

Multiplying (i) and (ii), we get

$$\bar{p}[1 - F(\bar{p})]^2 > \hat{p}[1 - F(\hat{p})]^2.$$

However, this is a contradiction, because $\hat{p}$ maximizes the right-hand side of the above inequality. Hence $\bar{p} \geq \hat{p}$.

APPENDIX E: PROOFS OF PROPOSITION 8 AND PROPOSITION 9

The proof proceeds by direct comparison between the seller's equilibrium expected payoff under incomplete information and the corresponding quantity under complete information; the latter is worked out in Lee (1994). To make the results of our model comparable to those of Lee's model, which assumes discounting over periods, we shall be dealing with the case in which the discount factor in Lee's model, δ, is equal to 1.

The intuition behind the result has already been alluded to in the text, namely that the seller finds it optimal to make lower offers on average because of her inability to verify the buyer's outside option and thus to match it, as she could under complete information. This intuition does not lead to a proof, since the preceding statement has to be verified by deriving both of the prices offered and the probability of acceptance by the buyer under the different search cost regimes, and this effectively leads to the comparison below. Note in particular that the incomplete information equilibrium depends on the endogenously derived cutoff b, which has no counterpart in the complete information model.

Following Lee (1994), when $\delta = 1$, the seller's equilibrium expected payoff under complete information, denoted as $S_c(p)$, is

$$S_c(p) = \max \begin{cases} p^* \\ \max_{p_1 \in [1 - \Phi(p''), 1 - \Phi(\bar{p})]} \displaystyle\int_{p^*}^{p''} p^* f(x_1) \, dx_1 \\ \qquad\qquad\qquad\qquad\qquad\qquad\qquad\qquad (i.a) \\ + \displaystyle\int_{p''}^{x^*(p_1)} x_1[1 - F(x_1)] f(x_1) \, dx_1 \qquad (i.b) \\ + \displaystyle\int_{x^*(p_1)}^{1} p_1 f(x_1) \, dx_1, \end{cases}$$

where $p'' \equiv \inf\{p \,|\, p[1 - F(p)] \geq p^*\}$, $x^*(p_1)$ satisfies $1 - p_1 = \Phi(x^*(p_1))$, and $\bar{p}$ maximizes $p[1 - F(p)]$.

Let $\tilde{p}_c$ be the maximizer and $S'_c(\tilde{p}_c)$ be the maximum value of $(i.b)$. Then $S_c(p)$ can be rewritten as

$$S_c(p) = \max\{p^*, S'_c(\tilde{p}_c)\}. \tag{ii}$$

In equilibrium, the seller offers p^* in the first period if $p^* \geq S'_c(\tilde{p}_c)$; otherwise, she offers $p_1 = \tilde{p}_c$.

Under incomplete information, the seller's equilibrium expected payoff, denoted as $S(p)$, is

$$S(p) = \max\begin{cases} p^* & (iii.a) \\ \max_{p_1 \in P_M}[F(b) - F(\hat{p}(b))]\,\hat{p}(b) \\ \quad \times [1 - F(\hat{p}(b))] + p_1[1 - F(b)], & (iii.b) \end{cases}$$

where P_M in $(iii.b)$ is defined as the set of all p_1 satisfying conditions of Proposition 4, and $\hat{p}(b)$ is the maximizer of $\Psi(p_2, b)$ defined in (3).

Now let $\tilde{p}$ maximize $(iii.b)$. Then $S(p)$ can be rewritten as

$$S(p) = \max\{p^*, [F(b) - F(\hat{p}(b))]\,\hat{p}(b)[1 - F(\hat{p}(b))]$$
$$+ \tilde{p}[1 - F(b)]\}. \tag{iv}$$

In equilibrium, the seller offers $p_1 = p^*$ if p^* is the maximum of (iv); otherwise, she offers $\tilde{p}$.

We want to show that $(i.b)$ is always greater than $(iii.b)$.

First, note that $\tilde{p}_c$, the maximizer of $(i.b)$, is chosen from the interval $[1 - \Phi(p''), 1 - \Phi(\bar{p})]$. We will now show, using the quasi-concavity of $p(1 - F(p))$, that $\tilde{p}$, the maximizer of $(iii.b)$, is also in the same interval. We then use the fact that $\tilde{p}_c$ maximizes $S_c(\cdot)$ within this interval to obtain an inequality.

Since $\tilde{p}_c = 1 - \Phi(x^*(\tilde{p}_c))$, the condition $\tilde{p}_c \in [1 - \Phi(p''), 1 - \Phi(\bar{p})]$ is equivalent to the condition that $x^*(\tilde{p}_c) \in [p'', \bar{p}]$. Furthermore, the equilibrium under the incomplete information model requires that $\tilde{p} = 1 - \Phi(\hat{p}(b))$. Hence, showing $\tilde{p} \in [1 - \Phi(p''), 1 - \Phi(\bar{p})]$ is equivalent to showing that $\hat{p}(b) \in [p'', \bar{p}]$.

Recall that $\bar{p}$ maximizes $p[1 - F(p)]$, and $\hat{p}(b)$ maximizes $\Psi(p_2, b)$. By Lemma 2 and Lemma 3, we know that $\bar{p} \geq \hat{p}(b)$ for all b.

We now show that $\hat{p}(b) \geq p''$.

We know that all $\hat{p}(b)$ satisfying the conditions of Proposition 4 must satisfy

$$\frac{F(b) - F(\hat{p}(b))}{F(b) - F(p^*)}\,\hat{p}(b)[1 - F(\hat{p}(b))] \geq p^* = p''[1 - F(p'')]. \tag{v}$$

Hence $\hat{p}(b)[1 - F(\hat{p}(b))] \geq p''[1 - F(p'')]$

Now, since $p[1 - F(p)]$ is quasi-concave and $\hat{p}(b)$ and p'' are both less than $\bar{p}$, we know that $\hat{p}(b) > p''$.

Hence we have shown that $\hat{p}(b)$ is in $[p'', \bar{p}]$, which implies that $\tilde{p} \in [1 - \Phi(p''), 1 - \Phi(\bar{p})]$.

Now, by the fact that $\tilde{p}_c$ is the maximizer of $S'_c(\tilde{p}_c)$, the following must be true:

$$S'_c(\tilde{p}_c) \tag{vi.a}$$

$$= \int_{p^*}^{p''} p^* f(x_1)\, dx_1 + \int_{p''}^{x^*(\tilde{p}_c)} x_1[1 - F(x_1)]f(x_1)\, dx_1$$

$$+ \int_{x^*(\tilde{p}_c)}^{1} \tilde{p}_c f(x_1)\, dx_1 \tag{vi.b}$$

$$> \int_{p^*}^{p''} p^* f(x_1)\, dx_1 + \int_{p''}^{\hat{p}(b)} x_1[1 - F(x_1)]f(x_1)\, dx_1$$

$$+ \int_{\hat{p}(b)}^{1} \tilde{p} f(x_1)\, dx_1 \tag{vi.c}$$

$$> \tilde{p}[1 - F(\hat{p}(b))] \tag{vi.d}$$

$$= \int_{\hat{p}(b)}^{b} \tilde{p} f(x_1)\, dx_1 + \int_{b}^{1} \tilde{p} f(x_1)\, dx_1 \tag{vi.e}$$

Note that $(iii.b)$ is equal to $\int_{\hat{p}(b)}^{b} \hat{p}(b)[1 - F(\hat{p}(b))]f(x_1)\, dx_1 + \int_{b}^{1} \tilde{p} f(x_1)\, dx_1$.

Since $\tilde{p} = 1 - \Phi(\hat{p}(b)) > \hat{p}(b)[1 - F(\hat{p}(b))]$, it is clear that the value of $(vi.e)$ must be greater that of $(iii.b)$. Hence, $(i.b)$ must be greater than $(iii.b)$. Therefore, $S_c(p)$, the seller's equilibrium expected payoff under complete information, is greater than or at least equal to $S(p)$, her equilibrium expected payoff under incomplete information.

Notice that, since $(i.b)$ is greater than $(iii.b)$, whenever it is optimal for the seller to offer the search-deterring price, p^*, in the complete information model (i.e., when $p^* \geq (i.b)$), it must also be optimal for the seller to offer p^* in the incomplete information model. However, the reverse is not true. Hence the seller will offer the search-deterring price for a greater range of c in the incomplete information model than in the complete information model. In other words, under incomplete information, it is easier for players to reach agreement immediately and hence achieve the efficient outcome. ∎

REFERENCES

Bester, H. (1988). "Bargaining, Search Costs and Equilibrium Price Distributions," *Review of Economic Studies* **55**, 201–214.

Binmore, K. (1985). "Bargaining and Coalitions," in Alvin E. Roth (ed.), *Game Theoretic Models of Bargaining*, Cambridge University Press.

Binmore, K. G., Shaked, A., and Sutton, J. (1989). "An Outside Option Experiment," *Quarterly Journal of Economics* **104**, 753–770.

Chatterjee, K., and Dutta, B. (1995). "Rubinstein Auctions: On Competition for Bargaining Partners," Mimeo, Pennsylvania State University and Indian Statistical Institute, Delhi.

Chatterjee, K., and Samuelson, W. (1983). "Bargaining under Incomplete Information," *Oper. Res.* **31**, 835–851.

Chatterjee, K. and Lee, C. C. (1996). "Bargaining and Search with Incomplete Information about Outside Options," Mimeo, Pennsylvania State University and the Chinese University of Hong Kong.

Chikte, S. D., and Deshmukh, S. D. (1987). "The Role of External Search in Bilateral Bargaining," *Oper. Res.* **35**, 198–205.

Farrell, J., and Gibbons, R. (1989). "Cheap Talk Can Matter in Bargaining," *J. Econom. Theory* **48**, 221–237.

Fudenberg, D., Levine, D., and Tirole, J. (1987). "Incomplete Information Bargaining with Outside Opportunities," *Quarterly Journal of Economics* **102**, 37–50.

Gresik, T. (1991). "The Efficiency of Linear Equilibria of Seal-Bid Double Auctions," *J. Econom. Theory* **53**, 173–184.

Harris, M. (1987). *Dynamic Economic Analysis*. Oxford: Oxford Univ. Press.

Lee, C. C. (1994). "Bargaining and Search with Recall: A Two-Period Model with Complete Information," *Oper. Res.* **42**, 1100–1109.

Muthoo, A. (1995). "On the Strategic Role of Outside Options in Bilateral Bargaining," *Oper. Res.* **43**, 292–297.

Myerson, R. B., and Satterthwaite, M. A. (1983). "Efficient Mechanisms for Bilateral Trading," *J. Econom. Theory* **29**, 265–281.

Raiffa, H. (1982). *The Art & Science of Negotiation*. Cambridge: Harvard Univ. Press.

Rubinstein, A. (1982). "Perfect Equilibrium in a Bargaining Model," *Econometrica* **50**, 97–109.

Satterthwaite, M. A., and Williams, S. R. (1989). "Bilateral Trade with the Sealed Bid K-Double Auction: Existence and Efficiency," *J. Econom. Theory* **48**, 107–133.

Shaked, A. (1987). "Opting Out: Bazaars versus 'High Tech' Markets," Discussion Paper 87/159 (Theoretical Economics), Suntory Toyota International Centre for Economics and Related Disciplines, London School of Economics.

Shaked, A., and Sutton, J. (1984). "Involuntary Unemployment as a Perfect Equilibrium in a Bargaining Model," *Econometrica* **52**, 1351–1364.

Wolinsky, A. (1987). "Matching, Search, and Bargaining," *J. Econom. Theory* **42**, 311–333.

GAMES AND ECONOMIC BEHAVIOR **23**, 119–145 (1998)
ARTICLE NO. GA970617

Rubinstein Auctions: On Competition for Bargaining Partners*

Kalyan Chatterjee

The Pennsylvania State University, University Park, Pennsylvania 16802

and

Bhaskar Dutta[†]

*Indian Statistical Institute, Delhi Centre, 7, S.J.S. Sansanwal Marg,
New Delhi, 110 016 India*

Received July 17, 1995

This paper analyzes the effect of competition for bargaining partners on the prices that prevail in thin markets, as well as how the matches are simultaneously determined. Three trading processes or bargaining procedures are described. In all the variants that we consider, except for one case of public offers, either there is no pure strategy subgame perfect equilibrium or such equilibria exhibit delay in reaching agreement. *Journal of Economic Literature* Classification Numbers: C72, D43.
© 1998 Academic Press

1. INTRODUCTION

A major lacuna in the traditional theory of competitive equilibrium is that the *process* by which the equilibrium price is *reached* is not really modelled. The seminal paper of Rubinstein (1982) has given birth to a rich literature which applies noncooperative bargaining theory to the study of the determination of prices.[1] Although the "equilibrium" price does not always

---

*We wish to thank the Penn State Center for Research in Conflict and Negotiation, the Center for Economic Research at Tilburg, the Indian Statistical Institute, and St. John's College, Cambridge, for hospitality during different phases of the work on this paper. Our greatest debt is to an anonymous referee, whose detailed and perceptive comments enabled us to detect an error in an earlier version of the paper. We are also grateful to Helmut Bester, Sudipto Bhattacharya, E. Hendon, and T. Tranaes for useful comments.
†To whom correspondence should be addressed. E-mail: dutta@isid.ernet.in.
[1]See Osborne and Rubinstein (1990) for an elegant and exhaustive description of this literature.

coincide with the competitive or market-clearing price in these strategic models of decentralized trade, it is generally the case that if agents are *perfectly informed* and if the good is *homogeneous*, then trade takes place at a *single price*, and without *delay*.[2]

In much of this literature, it is assumed that buyers and sellers are matched through some *random* procedure. An alternative assumption is that the order in which agents can make or respond to (price) offers is given *exogenously*. Clearly, while the assumption of a random matching procedure is an acceptable modelling device in large, anonymous markets (as, for example, in Gale, 1986, and Rubinstein and Wolinsky, 1985), it is less appropriate in thin markets where search costs are usually low. For similar reasons, the assumption of an exogenous order of offerors and responders is inappropriate in small markets, particularly when agents are heterogeneous since agents may then have an active interest in *choosing* their partners.

In this paper, we investigate the properties of some trading processes (or extensive forms) where the matching procedure is more appropriate for small markets with competition on both sides. We focus on the simplest, nontrivial extension of Rubinstein's original model. Our model consists of two identical sellers, each with one unit of an indivisible good, and two buyers with *different* valuations for the item. We consider three variants of an extensive form where agents on one side of the market, say buyers, make offers *simultaneously*, while sellers also respond *simultaneously* by either accepting or rejecting offers. Each offeror is allowed only one offer per period, while a responder can accept at most one offer. An acceptance means that the matched pair leaves the market. Unmatched pair(s) move on to the next period and interchange roles of offeror and responder, payoffs being discounted by a common discount factor. The three variants differ in terms of the nature of the offer and in the structure of information. Details are described in the next section.

Our procedures have the following advantages. First, a player *chooses* whom he or she will be matched with. Second, there is a minimal amount of order dependence since offers and responses are made simultaneously. Third, price offers either precede matching or are part of the same move, so a match has no "locking in" effect. Fourth, the procedure becomes a Rubinstein (1982) bargaining model with one seller and one buyer, and essentially an auction with one seller and two buyers. Our generalization therefore has some of the characteristics of both bidding and bargaining, as most real markets do. The three different trading processes we consider are appropriate for somewhat different institutions and give somewhat different results.

---

[2]An exception is Hendon and Tranaes (1991). We discuss their result shortly.

COMPETITION FOR BARGAINING PARTNERS 121

The main purpose of the paper is to analyse the effect of competition for bargaining partners on the price (or prices) that prevail in "thin" markets, as well as how the matches themselves are simultaneously determined. The matching literature[3] has concentrated primarily on co-operative notions of stability like the core, without discussing how the matches get made noncooperatively. We investigate this question, albeit in a simple version of the game. The extensive work on outside options or on bargaining in market contexts[4] has either looked at exogenous random matching of agents or exogenously given outside options or extensive forms that impose an order of proposers or responders. Many of these papers have dealt with one seller and two buyers, though not in the auction-like setting of this paper.[5]

The results we obtain are quite surprising. In all the variants that we consider here, either there is *no* pure strategy subgame perfect equilibrium[6] or such equilibria exhibit *delay* in reaching agreement,[7] thus ensuring that an efficient outcome is not obtained. It is worth emphasizing that typically simultaneous move games are characterized by "lots" of equilibria, and that the "delay" result is not due to any imperfect information. Most important, our results are at variance with the "received doctrine" of *efficient* trade (no delay in reaching agreement). Also, notice that since the main message of our paper is *negative*, not much purpose would be served by considering a more general model with a larger number of agents or by allowing more heterogeneity on both sides of the market.

The intuition behind these results is as follows. If both agreements take place in the same period, there is a tension between two forces on the price. The fact that a single rejection will generate a "Rubinstein bargaining subgame" leads to pressure on the prices to diverge towards the two bargaining solutions. However, the simultaneity of offers leads to competition and, therefore, undercutting or overbidding. This tends to push prices toward a value that is immune to undercutting. This tension cannot be resolved if two agreements are to occur in the same period.

---

[3]See Roth and Sotomayor (1990) and Shubik (1982), the latter for the "assignment game" that describes a situation very similar to ours.

[4]See Binmore (1985), Hendon and Tranaes (1991), Rubinstein and Wolinsky (1990), Shaked (1994), and the references in Osborne and Rubinstein (1990).

[5]Related work also appears in some recent papers on non-cooperative theories of coalitional bargaining. See, for example, Chatterjee *et al.* (1993), Bennett and van Damme (1991), Perry and Reny (1994), and Moldovanu and Winter (1992).

[6]We consider only pure-strategy equilibria until the discussion of private offers in Section 5.

[7]This is not quite accurate. See Proposition 2 and Remark 1, dealing with the public offers ("bazaar") trading process. Public offers is normally considered the most appropriate setting for core outcomes to emerge. In our case, the delay precludes this from happening in general, but the equilibrium prices tend to a single value as $\delta \to 1$, where δ is the common discount factor.

We also characterize the *unique* mixed strategy equilibrium that satisfies the property that a response decision depends only on the offer received by the particular responder. This equilibrium turns out to be intuitively more appealing.

The results of Hendon and Tranaes (1991) bear a (superficial) resemblance to ours. They consider the case of a single seller of valuation zero and two buyers B_1 and B_2 with valuations $v_1 > v_2 > v_1/2$.[8] In their paper, the seller is randomly matched with one of the buyers, and given the match, a random proposer is selected. Rejection of a proposal leads back to another random match. Hendon and Tranaes show that a stationary pure strategy equilibrium does not exist and that there is delay (no immediate agreement) if the seller is matched initially with the low value buyer. This occurs because there is a positive probability of meeting the high value buyer in the next period. However, if we consider our extensive form with one seller and two buyers, these results do not appear. The seller offers B_1 a price $v_1(1 - \delta) + \delta v_2$ (where δ is the discount factor) and this is accepted. If the buyers make offers, both offer v_2 and the seller accepts the high value buyer's offer. The possible delay in Hendon and Tranaes is an artefact of their random matching process. With two identical sellers and two heterogeneous buyers, however, delay arises in our model *without* random matching. Thus our model is qualitatively different from models like Hendon and Tranaes.

In Section 2, we describe the trading processes in greater detail. Sections 3–5 contain the main results.

2. THREE TRADING PROCESSES

The situation we model is the following. Two identical sellers S_1 and S_2 each own one unit of an indivisible good. For simplicity, we assume that both sellers have the same reservation value of zero for the good. There are also two buyers B_1 and B_2, both of whom demand one unit each of the commodity. The buyers' valuations are v_1 and v_2, respectively, with $v_1 > v_2 > v_1/2 > 0$. Valuations are common knowledge.

The price(s) at which the good is exchanged if trade takes place is determined by *bargaining* among the agents. In this paper, we describe three alternative bargaining procedures. We call the first two the *public offers* and *targetted* offers models, while the third is the *telephone bargaining* model of Binmore (1985). Telephone conversations in our model are, however, pri-

---

[8]We also make this assumption on the valuations in order to make it possible for B_2 to bid up the price above the "Rubinstein" price for B_1.

vate, unlike in Binmore (1985), which also differs from us in being about three players and about sequential rather than simultaneous offers.

First, we describe the public offers model. In each period $t \in \{1, 2, \ldots\}$, agents (say, buyers) on one side of the market *simultaneously* announce a price p_i at which they are willing to buy one unit of the good. The two sellers then respond (again *simultaneously*) to the price offers. A response is either acceptance of *one* offer or rejection of both offers. If *both* sellers accept B_i's offer of p_i, then the two sellers are *matched* with equal probability with B_i. Both pairs are matched if the two sellers accept offers from different buyers. Matched pairs leave the market with the good being exchanged at the agreed price offer.

If some pair(s) remain unmatched at the end of period t, then in period $(t + 1)$ the game is repeated with sellers making price offers and buyers responding to these offers, agents on both sides of the market moving simultaneously as in period t. This procedure is repeated so long as some pair remains in the market. All agents have a common discount factor $\delta \in (0, 1)$.

The *targetted* offers model is very similar. The only difference is that, unlike in the public offers model, an *offer* consists of a price p *and* an agent on the other side of the market. Thus, when buyers make offers, B_i names a price p_i along with a seller S_j with whom B_i wants to trade. And only seller S_j can accept or reject the offer of price p_i from B_i. Also, even if two offers are received by a particular agent, the agent can accept at most one offer. As in the public offers model, matched pair(s) leave the market, while unmatched pair(s) continue the bargaining process with buyers and sellers interchanging roles of offerors and responders.

In the targetted offers model, all offers are "heard" by both responders. Thus, even if (say) B_i makes an offer to S_i and B_j makes an offer to S_j, S_i and S_j are aware of both offers. So, in principle, S_i's response decision can also depend on the offer received by S_j. The telephone bargaining model is a variant of the targetted offers model in which offers are "transmitted" through telephone calls. In other words, if B_i makes an offer to S_i and B_j to S_j, then S_i is unaware of the offer received by S_j and conversely. This implies that response decisions can only depend on the offer(s) received by oneself.

Each of these corresponds to a different market institution. Many markets are characterized by sellers fixed in different locations, so that a buyer offer to one is not "heard" by others. These are characteristic of "telephone bargaining," and the car and housing markets are obvious examples. Targetted offers are less common. One example which comes to mind is the market for takeover targets. Buyers make public offers, and the responses following a rejection are also public. Merger negotiations correspond more to "telephone bargaining." Finally, note that the public offers model corresponds more to the "bazaars" of the world, in which there are posted prices which can be negotiated down. It could also be thought of as a model of prices being called out in stock markets.

Clearly, the bargaining processes described above can all be viewed as natural generalizations of the alternating offers model of Rubinstein (1982). Our purpose is to analyse the subgame perfect equilibria of these processes for "high" discount factors. In Sections 3 and 4, we focus on *pure strategy* equilibria of the public offers and targetted offers models, respectively. We characterize the *unique* subgame perfect pure strategy equilibrium in the public offers model. This equilibrium exhibits *delay* (that is, in any period t, only one pair is matched in equilibrium) when buyers make offers. In the targetted offers model, all pure strategy equilibria *must* involve *delay*. Moreover, the pure strategy equilibria of the targetted offers model can only be sustained by response decisions which depend on the offer received by the other agent. This implies that there is *no* pure strategy equilibrium in the telephone bargaining model. This prompts us to look at *mixed-strategy* equilibria in the telephone bargaining model. In Section 5, we show that there is a unique mixed-strategy equilibrium in the model.

In the remainder of this section, we will prove some preliminary results which will be useful in the analysis of pure strategy equilibria in both the public offers and targetted offers models.

Henceforth, we will denote by S-games (B-games) the (sub)games in which sellers (respectively, buyers) make offers. Payoffs will be represented by $R(B_i)$, $R(S_j)$, etc. Thus, if one unit of the good is exchanged in period t between B_i and S_j at price p, then $R(B_i) = \delta^{t-1}(v_i - p)$ and $R(S_j) = \delta^{t-1}p$.

First, we note that equilibrium cannot entail bargaining forever. For, if it did, B_1 could deviate and offer a price δv_1, which would be accepted by either seller, thus giving a positive payoff to the deviator.

If an equilibrium exists therefore, it must consist of two agreements, either both in period t or in periods $(t, t+1)$. Also, note that if only one pair reaches agreement in period t, then the remaining pair will be engaged in a Rubinstein alternating offers subgame. We will often refer to *S-equilibrium* and *B-equilibrium* payoffs. An S-equilibrium payoff will mean an equilibrium in which *at least one seller's offer is accepted*, the remaining seller (if any) being then engaged in the Rubinstein alternating offers subgame in the next period. A similar interpretation is to be given to B-equilibrium payoffs.[9]

In checking possible equilibria in Sections 3 and 4, we will often need to consider the case where a deviation from his or her equilibrium strategy by one proposer in, say, the B-game leads to *both* responders rejecting the offers. The rejection decisions must then be sustained by equilibrium pay-

---

[9]We occasionally refer to these also as S-game and B-game equilibrium payoffs.

offs in the S-game. We prove this formally for the B-game of the targetted offers model. The reader can check that the result remains true in other cases.

Let Q be the set of possible equilibrium *payoffs* (b_1, b_2, s_1, s_2) of the B-game in the targetted offers model which are supported by strategies satisfying the following:

(i)   There exists a deviation p_i for B_i such that $v_i - p_i > b_i$.

(ii) S_i accepts p_i if S_j, $j \neq i$, accepts p_j, the offer from B_j.

Note that, in general, Q is a *subset* of the set of possible equilibrium payofffs.

LEMMA 0.   *Suppose $(b_1, b_2, s_1, s_2) \in Q$. Then there is an equilibrium in the S -game with seller payoffs (m_1, m_2) such that $(s_1, s_2) \leq \delta(m_1, m_2)$.*

See the Appendix for the proof.

Remark 1.   What Lemma 0 asserts is the following. Consider a strategy profile for the players and suppose it implies that at least one agreement occurs in the B-game. In order to check whether this is an equilibrium, we need to consider the four-player subgame resulting from two rejections. Then it suffices to check only the possible equilibrium continuation payoffs involving at least one agreement in the S-game. In particular, we do not require to check the situation which follows from two rejections in the S-game.

3. THE PUBLIC OFFERS MODEL

In this section, we focus attention on the public offers model. The main results, Propositions 1 and 2, show that there is a *unique* subgame perfect equilibrium in pure strategies. The nature of the equilibrium differs between the B-game and the S-game. In the B-game, the equilibrium involves only *one* agreement in the initial period. In the S-game equilibrium, both sellers charge the same price and both buyers accept this price. Thus, the nature of the equilibrium appears to depend on the heterogeneity of buyers.

A couple of lemmas precede the main propositions.

LEMMA 1.   *In the public offers model, any possible equilibrium payoff vectors of sellers, (s_1, s_2), in the S-game must satisfy one of the following restrictions:*

(a) $(s_1, s_2) \leq [v_1/(1 + \delta), v_2/(1 + \delta)]$.

(b) $(s_1, s_2) \leq [v_2/(1 + \delta), \frac{1}{2}\delta^2(v_1 + v_2)/(1 + \delta)]$.

(c) $(s_1, s_2) = [p_1, \delta^2 v_2/(1 + \delta)]$, *where* $p_1 \geq v_2/(1 + \delta)$.

(d) $(s_1, s_2) = [p_2, \delta^2 v_1/(1 + \delta)]$, *where* $p_2 \geq v_1/(1 + \delta)$.

Proof. Suppose both pairs are matched in the S-game. Then, B_i can by rejecting the offers precipitate a Rubinstein subgame. This gives B_i a payoff of $v_i/(1 + \delta)$ in the *next* period. Hence, $R(B_i) \geq \delta v_i/(1 + \delta)$, so that $p_i \leq v_i/(1 + \delta)$. This establishes (a).

If both buyers accept an offer p from the *same* seller S_i, then B_2 has (expected) payoff $R(B_2) = \frac{1}{2}[(v_2 - p) + \delta v_2/(1 + \delta)]$. Rejection of both offers gives B_2 a payoff of $\delta v_2/(1 + \delta)$. Hence, $p \leq v_2/(1 + \delta)$. Also, seller S_j $(j \neq i)$ is matched with either buyer with equal probability in a Rubinstein subgame. This gives the seller a payoff of $\frac{1}{2}\delta^2(v_1 + v_2)/(1 + \delta)$. This proves (b).

Suppose now that B_i rejects *both* offers while B_j accepts p_j from S_j. Then, $R(B_i) = \delta v_i/(1 + \delta)$, whereas acceptance of p_j would have given B_i an expected payoff of $\frac{1}{2}[(v_i - p_j) + \delta v_i/(1 + \delta)]$. So, $\delta v_i/(1 + \delta) \geq v_i - p_j$ to sustain rejection of p_j as an equilibrium. Also, the unmatched seller S_i gets $\delta^2 v_i/(1 + \delta)$. These yield (c) and (d).   ∎

LEMMA 2.   *In the public offers model, any possible equilibrium payoff vectors of buyers, (b_1, b_2), in the B-game must satisfy one of the following restrictions*:

(a) $(b_1, b_2) \leq [v_1/(1 + \delta), v_2/(1 + \delta)]$.

(b) $(b_i, b_j) \leq [v_i - p_i, \delta^2 v_j/(1 + \delta)]$ *for* $i, j = 1, 2$, *where* $p_i \geq \delta v_j/(1 + \delta)$.

(c) $(b_i, b_j) = [v_i - p_i, \delta^2 v_j/(1 + \delta)]$ *for* $i, j = 1, 2$, *where* $p_i \leq \delta v_j/(1 + \delta)$.

Proof. The proof of (a) is analogous to that of Lemma 1(a), with the upper bounds on buyer payoffs following from the sellers' payoffs in respective Rubinstein subgames. (See the Appendix.)

Suppose now that both sellers accept p_i from B_i in equilibrium. Then each seller has expected payoff of $R(S_i) = \frac{1}{2}[p_i + \delta v_j/(1 + \delta)]$. Rejection of both offers gives either seller a payoff of $\delta v_j/(1 + \delta)$. Hence, $p_i \geq \delta v_j/(1 + \delta)$. Also, B_j is in a Rubinstein subgame with the seller making the first offer. This establishes (b).

Case (c) corresponds to the case where one seller accepts p_i from B_i, while the other seller rejects both offers. The details are omitted.   ∎

We are now ready for the first main result of this section.

PROPOSITION 1. *For δ sufficiently high, there is a unique pure strategy equilibrium in the B-game of the public offers model.*[10]

Proof. The proposition is proved in several steps.

Step 1. Both sellers cannot accept B_2's offer. If both sellers *do* accept p_2 from B_2, then $p_2 \geq \delta v_1/(1 + \delta)$ from Lemma 2. Let B_2 deviate and offer $p'_2 = p_2 - \epsilon$, where $\epsilon > 0$ and $\delta^2 v_2/(1 + \delta) > v_2 - p'_2$. Note that such p'_2 exists for δ high enough.

To sustain the original equilibrium, *both* sellers must reject *both* offers since acceptance of p_1 will give B_2 a payoff of $\delta^2 v_2/(1 + \delta) > v_2 - p_2$. But, if both sellers reject p'_2, then there is a continuation payoff (s_1, s_2) in the S-game such that $\delta(s_1, s_2) \geq (\delta v_1/(1 + \delta), \delta v_1/(1 + \delta))$. From Lemma 1, it is easy to check that this is not true. Hence, Step 1 is proved.

Step 2. Both buyers cannot accept an offer from the same seller in the S-game. Suppose B_1 and B_2 both accept p_j from S_j. Now, if $p_i = p_j = p$, then B_1 will deviate and accept p from S_i. This will give B_1 a payoff of $(v_1 - p)$, whereas acceptance of p from S_j gives $\frac{1}{2}[v_1 - p + \delta v_1/(1 + \delta)] < v_1 - p$ since $p \leq v_2/(1 + \delta) < v_1/(1 + \delta)$. So, $p_i > p_j$.

Let S_j raise the offer "slightly" to $p'_j < p_i$. To sustain the original equilibrium, both buyers reject any such p'_j. Hence, there is a continuation payoff (b_1, b_2) in the B-game such that $\delta(b_1, b_2) \geq (v_1 - v_2/(1 + \delta), \delta v_2/(1 + \delta))$. Lemma 2 and Step 1 rule out this possibility.

Step 3. In the S-game, if B_i rejects both offers, then B_j must also reject both offers. Suppose first that B_2 accepts an offer whereas B_1 rejects both offers. Then

$$R(B_2) \leq v_2 - \frac{v_1}{(1 + \delta)} < \delta\left(v_2 - \frac{\delta^3 v_1}{1 + \delta}\right) \leq \delta \underline{b}_2, \tag{1}$$

where $\underline{b}_2$ is the minimum payoff of B_2 in the B-game. To check the last inequality in (1), notice that if both buyers' offers are accepted in the B-game, then $p_2 \leq \delta^3 v_1/(1 + \delta)$. For, suppose $p_2 > \delta^3 v_1/(1 + \delta)$. Let B_2 *reduce* the offer to $p'_2 = p_2 - \epsilon > \delta^3 v_1/(1 + \delta)$. Since $p_1 \geq \delta v_1/(1 + \delta)$, B_2's offer can be rejected only if there is an S-game continuation payoff $(\tilde{s}_1, \tilde{s}_2)$ such that $\delta(\tilde{s}_1, \tilde{s}_2) \geq (\delta v_1/1 + \delta, \delta^3 v_1/(1 + \delta) + \epsilon)$. But, since there is no such payoff vector, $p_2 \leq \delta^3 v_1/(1 + \delta)$. So, $\underline{b}_2 \geq v_2 - \delta^3 v_1/(1 + \delta)$, as was to be shown.

But, (1) establishes that if B_1 rejects both offers, then B_2 must also reject both offers.

---

[10]Since the proof proceeds constructively, the description of the equilibrium is contained in the body of the proof.

Suppose now that B_1 accepts an offer of $p \geq v_2/(1 + \delta)$, whereas B_2 rejects both offers. Letting S_j denote the unmatched seller, $R(S_j) = \delta^2 v_2/(1 + \delta)$. Let S_2 deviate and make an offer of $\delta^2 v_2/(1 + \delta) + \epsilon$, where $\epsilon > 0$. Then, to sustain the original equilibrium, both buyers reject both offers. But, this requires buyer continuation payoffs of $(v_1 - \delta^2 v_2/(1 + \delta), v_2 - \delta^2 v_2/(1 + \delta))$. But, no such payoffs are possible, thereby proving Step 3.

Step 4. B_2's offer cannot be accepted in equilibrium. In Step 1, we have ruled out both sellers accepting an offer from B_2. We now rule out the other possibilities which are consistent with B_2's offer being accepted in equilibrium.

First, both buyers' offers of p_1 and p_2 cannot be accepted. If both buyer offers are indeed accepted, then $p_2 \leq v_2(1 - \delta^2/(1 + \delta))$. For, if $p_2 > v_2(1 - \delta^2/(1 + \delta))$, then B_2 can deviate and offer $p_2' = v_2(1 - \delta^2/(1 + \delta))$. Since $p_1 \geq \delta v_1/(1 + \delta)$, and since rejection of p_2' implies rejection of p_1, this requires seller continuation payoffs (s_1, s_2) such that $\delta(s_1, s_2) \geq (\delta v_1/(1 + \delta), v_2(1 - \delta^2/(1 + \delta)))$. This is impossible. So, $p_2 \leq v_2(1 - \delta^2/(1 + \delta))$.

Let S_2 now reject B_2 and accept B_1's offer of p_1. Then

$$R(S_2) \geq \frac{1}{2}\left[\frac{\delta v_1}{1 + \delta} + \frac{\delta v_2}{1 + \delta}\right] > v_2\left(1 - \frac{\delta^2}{1 + \delta}\right)$$

for δ sufficiently close to 1. Hence, both buyer offers cannot be accepted.

Next, note that if one seller rejects both offers, while S_i accepts B_2's offer, then $R(B_1) = \delta^2 v_1/(1 + \delta)$. B_1 can deviate to a price offer of $p = v_1(1 - \delta^2/(1 + \delta)) - \epsilon$. For ϵ sufficiently small, $p > \delta v_1/(1 + \delta) = \delta \bar{s}$, where $\bar{s}$ is the maximum continuation payoff in the S-game. So, B_1's offer cannot be rejected. This completes the proof of Step 4.

In view of Steps 2 and 3, the only possible equilibrium in the S-game is for both seller offers to be accepted (Case (a) of Lemma 1). In Steps 1 and 4, we have already ruled out B_2's offer being accepted in equilibrium. The only remaining possibility is that B_1's offer is accepted in equilibrium. We now demonstrate that, for δ sufficiently high, there is a unique equilibrium with *both* sellers accepting B_1's offer.

Step 5. The following describes the equilibrium. Let p_1 and p_2 be the prices offered by B_1 and B_2, respectively. Then, let p_1, p_2 satisfy

$$\frac{1}{2}\left(p_1 + \frac{\delta v_2}{1 + \delta}\right) = v_2\left(1 - \frac{\delta^2}{1 + \delta}\right), \tag{2}$$

$$p_2 = v_2\left(1 - \frac{\delta^2}{1 + \delta}\right). \tag{3}$$

COMPETITION FOR BARGAINING PARTNERS   129

Sellers' response strategies to offers x_1 from B_1 and x_2 from B_2 are as follows:

(i)   Reject any offer x_1 if $x_1 < \delta v_2/(1+\delta)$.

(ii)   If $x_1 \geq \delta v_2/(1+\delta) > x_2$, accept x_1.

(iii)   If $x_1 < \delta v_2/(1+\delta) \leq x_2$, accept x_2 if $x_2 \geq \delta v_1/(1+\delta)$. If the latter inequality does not hold, then S_1 rejects both offers, while S_2 accepts x_2.

(iv)   If $x_1, x_2 \geq \delta v_2/(1+\delta)$, then

1.   both sellers accept x_1 if $\frac{1}{2}(x_1 + \delta v_2/(1+\delta)) \geq x_2$;

2.   both sellers accept x_2 if $x_2 > \frac{1}{2}(x_1 + \delta v_2/(1+\delta))$ and $\frac{1}{2}(x_2 + \delta v_1/(1+\delta)) \geq x_1$;

3.   if neither (1) nor (2) above holds, then S_2 accepts x_2 and S_1 accepts x_1 iff $x_1 \geq \delta v_1/(1+\delta)$.

We check that these strategies constitute an equilibrium with B_1's offer being accepted.

First, note that since $v_2(1 - \delta^2/(1+\delta)) > \delta v_2/(1+\delta)$, it is optimal for both sellers to accept B_1's offer of p_1. In this case, B_2 is matched in the next period with one of the sellers and gets a (discounted) payoff of $\delta^2 v_2/(1+\delta)$. Both sellers' expected payoffs equal $v_2(1 - \delta^2/(1+\delta))$. If B_2 deviates and offers a price of $x_2 = p_2 + \epsilon$, then the offer will be accepted by at least one seller. This gives B_2 a lower payoff. If B_2 offers a price that is lower than p_2, then the offer is rejected. So, B_2 does not have any profitable deviation.

If B_1 deviates and makes a *higher* price offer, then the offer will be accepted and B_1 will be worse off. If B_1 deviates to $p_1 - \epsilon$, then we shall show below that both sellers accept B_2's offer of p_2. The value of p_1 from (2) is calculated to be $2v_2(1-\delta) + \delta v_2/(1+\delta)$. Hence,

$$\frac{1}{2}\left(p_2 + \frac{\delta v_1}{1+\delta}\right) - p_1$$

$$= \frac{1}{2}\left(v_2\left(1 - \frac{\delta^2}{1+\delta}\right)\right) + \frac{1}{2}\frac{\delta v_1}{1+\delta} - 2v_2(1-\delta) - \frac{\delta v_2}{1+\delta}$$

$$= \frac{1}{2(1+\delta)}\left[v_2(3\delta^2 - 3 - \delta) + \delta v_1\right].$$

The expression in square brackets is increasing in δ for $\delta \in [0, 1]$ and is positive for $\delta = 1$. Therefore there is some value $\bar{\delta}$, such that, for $\delta > \bar{\delta}$, the expression is positive.[11]

So, B_1's (expected) payoff following this deviation is $\delta^2 v_1/(1 + \delta) < v_1 - p_1$ for δ sufficiently high. So a lower offer for B_1 is not profitable.

This completes the proof of the proposition. ■

Proposition 1 leaves open the possibility of an equilibrium in the S-game with *both* pairs being matched. We now show that there is indeed a *unique* pure strategy equilibrium in the S-game where both are matched at a price of $p_2 = v_2/(1 + \delta)$.

PROPOSITION 2. *The unique pure strategy equilibrium of the S-game in the public offers model involves both sellers offering a price $p^* = v_2/(1 + \delta)$, and both pairs are matched.*

Proof.   We first show that this is indeed an equilibrium.

Let S_1 and S_2 both offer $p^* = v_2/(1 + \delta)$.

B_i accepts offer from S_i for $i = 1, 2$. Also, for any vector (p_i, p_j) with $p_i = p^*$ and $p_j > p^*$, B_2 rejects both offers, while B_1 accepts p_i. (This is, of course, not a complete description of buyers' responses.)

Then it is trivial to check that these strategies constitute an equilibrium.

We now show why this is the only equilibrium in the S-game. Let $(\hat{p}_1, \hat{p}_2)$ be equilibrium offers. From Lemma 1(a), $\hat{p}_i \leq v_i/(1 + \delta)$. So, the only other possibility is that $\hat{p}_1 \neq \hat{p}_2 \leq v_2/(1 + \delta)$, where $\hat{p}_i$ is the price accepted by B_i, $i = 1, 2$. But then the seller offering $\hat{p}_2$ will deviate and *increase* the offer to $p'_2 = \hat{p}_1 - \epsilon$. B_1 cannot reject this offer, and so $(\hat{p}_1, \hat{p}_2)$ cannot be equilibrium offers. ■

Remark 2.   It is worth pointing out that even a small amount of heterogeneity among sellers will destroy this equilibrium.

Remark 3.   The preceding propositions have not considered the case where *both* offers, say by a buyer, are rejected in equilibrium. Suppose that the equilibrium payoffs of the players in such an equilibrium are $\delta(b_1, b_2, s_1, s_2)$, that is, the rejection leads to two agreements in the S-game. Then buyer 1 can improve his payoff by offering the seller with whom he is eventually matched a price of $\delta v_2/(1 + \delta)$ plus ϵ. The seller concerned will only reject, given the other seller continues to reject, if the ensuing

----

[11]The value of $\bar{\delta}$ must satisfy

$$\frac{v_1}{v_2} = \frac{3 + \bar{\delta} - 3\bar{\delta}^2}{\bar{\delta}}.$$

COMPETITION FOR BARGAINING PARTNERS   131

four-player S-subgame will generate a payoff greater than this. However, the S-equilibrium payoff for a seller is unique; therefore the greater payoff must be generated by a B-equilibrium payoff for the seller. But the sellers get equal payoffs in the only B-equilibrium and this payoff (discounted by δ^2) can be checked to be smaller than $\delta v_2/(1 + \delta)$, and therefore not capable of sustaining the original seller rejections. An analogous argument, based on the uniqueness of the agreement payoffs, shows that two buyer rejections cannot be in equilibrium.

4. THE TARGETTED OFFERS MODEL

In this section, we analyse the nature of pure strategy equilibria in the targetted offers model. The principal result of this section shows that only *one* pair of agents can be matched in the first period of any pure strategy equilibrium, the other pair of agents being matched in the next period. This is true irrespective of whether buyers or sellers make offers. It is worth pointing out that the *inefficiency* arising out of delay in reaching agreement is not the result of "wrong" matchings arising from random matchings (as in Hendon and Tranaes, 1991), or incomplete information.

Let $\underline{\delta}$ satisfy the following inequality:

$$\frac{v_2}{v_1} \le \frac{\delta^2}{1 + \delta - \delta^3}. \tag{4}$$

The results in this section hold for all $\delta \ge \underline{\delta}$. The next lemma is proved in the Appendix. Let $\bar{s}$ denote the maximum seller payoff in the S-game of the targetted offers model.

LEMMA 3.   *If $\delta \ge \underline{\delta}$, then $\bar{s} \le v_1/(1 + \delta)$.*

Proof.   See the appendix for the proofs.

Table I describes the possible equilibrium configurations in the targetted offers model.

PROPOSITION 3.   *If $\delta \ge \underline{\delta}$, every pure strategy equilibrium in the targetted offers model must exhibit delay.*

Proof.   The proof is broken up into several steps.

Step 1.   Case 6 cannot occur in equilibrium. In view of Lemma 3, S_1 cannot reject any offer higher than $\delta v_1/(1 + \delta)$. So, B_1 can deviate and make an acceptable offer of $\delta v_1/(1 + \delta) + \epsilon$.

TABLE I

Possible Equilibrium Configurations in the Targetted Offers Model[a]

Case 1. Both seller offers accepted in equilibrium. Then $p_i \leq v_i/(1 + \delta)$, where p_i for
$i = 1, 2$ is the price agreed by (B_i, S_i).

Case 2. S_1 offers p_1^s to B_1, who accepts. S_2 and B_2 do not have an agreement but agree
in the two-player game with B_2 as proposer. Here

$$R(S_2) = \frac{\delta^2 v_2}{1 + \delta} \qquad R(B_2) = \frac{\delta v_2}{1 + \delta}.$$

Case 3. S_2 offers p_2^s to B_2, who accepts. B_1 and S_1 agree in the two-player subgame with
B_1 as proposer. Here

$$R(S_1) = \frac{\delta^2 v_1}{1 + \delta} \qquad R(B_1) = \frac{\delta v_1}{1 + \delta}.$$

Case 4. Both pairs reach agreement in the same period in the B-game. Then $p_i \geq$
$\delta v_i/(1 + \delta)$, where p_i is the price agreed by (B_i, S_i) for $i = 1, 2$.

Case 5. B_1 offers p_1^B to S_1, who accepts. B_2 and S_2 agree in the two-player subgame with
S_2 as proposer. Here

$$R(B_2) = \frac{\delta^2 v_2}{1 + \delta} \qquad R(S_2) = \frac{\delta v_2}{1 + \delta}.$$

Case 6. B_2 offers p_2^B to S_2, who accepts. B_1 and S_1 agree in the S_1-subgame. Here

$$R(B_1) = \frac{\delta^2 v_1}{1 + \delta} \qquad R(S_1) = \frac{\delta v_1}{1 + \delta}.$$

---

[a]Cases 1–3 pertain to the S-game, Cases 4–6 to the B-game. We have assumed, w.l.o.g., that
matches take place between B_i and S_i.

Step 2. Case 2 cannot occur in equilibrium. Note that from Cases 4 and
5, $\bar{b}_2 \leq v_2/(1 + \delta)$, where $\bar{b}_2$ is the largest equilibrium payoff of B_2 in
the B-game. Hence, S_2 can deviate and make an offer of $\delta v_2/(1 + \delta) -$
ϵ to B_2. Since B_2 must accept this offer, Case 2 cannot be an equilibrium.

Step 3. If both pairs are matched in equilibrium without delay, then
$p_i = \delta v_i/(1 + \delta)$ in the B-game, and $p_i = v_i/(1 + \delta)$ in the S-game.

Consider the B-game. We know that $p_i \geq \delta v_i/(1 + \delta)$, since a rejection
by one seller induces the Rubinstein subgame. Suppose $p_i > \delta v_i/(1 + \delta)$.
Let B_i deviate and offer S_i a price p^* such that $p_i > p^* > \delta v_i/(1 + \delta)$.
S_i can reject this offer only if S_j rejects p_j. This implies that there is a continuation seller payoff (s_1, s_2) in the S-game such that $\delta(s_1, s_2) \geq (p^*, p_j)$.
This cannot happen in Cases 1 and 3, while Case 2 has been ruled out in
Step 2. So, $p_i = \delta v_i/(1 + \delta)$.

The proof that $p_i = v_i/(1 + \delta)$ in the S-game is analogous.

Step 4. $p_1^B \geq \delta v_2/(1+\delta)$. Let $\underline{s}$ denote the minimum seller payoff in the S-game. Since a seller accepts p_1^B, we must have

$$p_1^B \geq \delta \underline{s}$$

$$\geq \delta \min\left(\frac{v_2}{1+\delta}, \frac{\delta^2}{1+\delta}v_1, p_2^s\right) \tag{5}$$

$$\geq \delta \min\left(\frac{v_2}{1+\delta}, p_2^s\right)$$

since $\delta^2 > v_2/v_1$.

Consider Case 3 of Table I. If S_1 offers a price $p = \delta^2 v_1/(1+\delta) + \epsilon$ to B_1, then B_1 must accept unless B_2 also rejects the offer from S_2. So, there is a continuation payoff vector (b_1, b_2) such that $\delta(b_1, b_2) \geq (v_1(1 - \delta^2/(1+\delta)), v_2 - p_2^s)$. The only possibility is that (b_1, b_2) is obtained from Case 5, where $R(B_2) = \delta^2 v_2/(1+\delta)$. Hence, $p_2^s \geq v_2(1 - \delta^3/(1+\delta)) \geq v_2/(1+\delta)$. From (5), $p_1^B \geq \delta v_2/(1+\delta)$.

Step 5. Both pairs cannot agree in equilibrium in the S-game. Suppose (S_i, B_i) are matched for $i = 1, 2$. Then we must have $p_i = v_i/(1+\delta)$ from Step 3.

Let S_2 deviate and offer $p = v_2/(1+\delta) + \epsilon$, where ϵ is a small, positive number. Then B_1 must reject p. Hence, there is an equilibrium payoff b_1 in the B-game such that $\delta b_1 \geq v_1 - v_2/(1+\delta)$. But, since $p_1^B \geq \delta v_2/(1+\delta)$, and $p_1 = \delta v_1/(1+\delta)$ if both pairs are matched in the B-game, there is no such equilibrium payoff for B_1.

Hence, both pairs cannot be matched in equilibrium.

Step 6. Both pairs cannot be matched in equilibrium in the B-game. If they are matched, then $p_i = \delta v_i/(1+\delta)$. Let B_1 deviate and offer a price arbitrarily smaller than $\delta v_1/(1+\delta)$ to S_2. S_2 must reject both offers. This implies that $\delta \bar{s} \geq \delta v_1/(1+\delta)$. But, since the only possible equilibrium in the S-game is Case 3, $\bar{s} \leq \max(p_2^s, \delta^2 v_1/(1+\delta))$. Since B_2 accepts p_2^s, $v_2 - p_2^s \geq \delta \underline{b}_2$, where $\underline{b}_2$ is the minimum equilibrium payoff of B_2 in the B-game. So $\underline{b}_2 = \delta^2 v_2/(1+\delta)$ (from Case 5), and $p_2^s \leq v_2(1 - \delta^3/(1+\delta))$. For $\delta \geq \bar{\delta}$, $\delta^2 v_1/(1+\delta) \geq v_2(1 - \delta^3/(1+\delta))$, and hence $\bar{s} \leq \delta^2 v_1/(1+\delta)$.

So, S_2 cannot reject an offer of $(\delta v_1/(1+\delta)) - \epsilon$.

This completes the proof of Proposition 3. ∎

Proposition 3 does not rule out the existence of pure strategy equilibria. Notice, however, that in the process of proving Proposition 3, we have ruled out Cases 1, 2, 4, and 6. Therefore, the equilibria, if any, must be in Cases 3 and 5. In the S-game, B_2 is in the first agreement. In the B-game, B_1 agrees first.

Let $\hat{\delta}$ satisfy the following inequality:

$$\frac{v_2}{v_1} \leq \min\left[\frac{\delta^3}{1 + \delta - \delta^2}, \frac{\delta(1 + \delta) - 1}{\delta^2(1 + \delta - \delta^3)}\right]. \tag{6}$$

Note that $\hat{\delta} > \underline{\delta}$.

PROPOSITION 4. *For $\delta \geq \hat{\delta}$, the following describes a pure strategy equilibrium configuration in the targetted offers model:*

S-game. S_2 proposes $p_2^s = v_2(1 - \delta^3/(1 + \delta))$ to B_2, who accepts. S_1 asks for v_1 from (or does not make an offer to) B_1, who rejects. After S_2 and B_2 have left the market, B_1 offers a price of $\delta v_1/(1 + \delta)$ to S_1, who accepts.

B-game. B_1 proposes $p_1^B = \delta v_2(1 - \delta^3/(1 + \delta))$ to S_2, who accepts. B_2 and S_1 do not agree until after (B_1, S_2) have left the market, when S_1 offers $v_2/(1 + \delta)$ to B_2, who accepts.

Proof. This is sustained by the following equilibrium strategies ("acceptance" implies acceptance of the best possible offer that satisfies the condition):

(i) S_2 proposes p_2^s to B_2. S_1 makes an irrelevant offer, as explained above.

(ii) B_2 accepts $p \leq p_2^s$ from S_2 if S_1 either makes no offer or a "non-serious" offer to B_1. If S_1 offers $p \leq \delta v_1/(1 + \delta)$ to B_1, B_2 rejects p_2^s but accepts $p < p_2^s$.

(iii) B_1 accepts p such that $v_1 - p \geq \delta[v_1 - \delta v_2(1 - \delta^3/(1 + \delta))]$.

(iv) In the B-game, B_1 proposes p_1^B to S_2, where $p_1^B = \delta v_2(1 - \delta^3/(1 + \delta))$. B_2 proposes 0.

(v) S_2 accepts offers $p \geq p_1^B$ unless B_2 proposes a price $p \geq v_1/(1 + \delta)$, when S_2 rejects offers $p \leq p_1^B$.

(vi) S_1 accepts the highest price above $\delta^3 v_1/(1 + \delta)$.

These constitute our equilibrium provided $\delta \geq \hat{\delta}$. Note that, in (iii), B_1 must reject the offer of $v_1/(1 + \delta)$ if S_1 deviates and offers this. Given that B_2 also rejects S_2's offer, B_1's payoff is then

$$\delta\left(v_1 - \delta v_2\left(1 - \frac{\delta^3}{1 + \delta}\right)\right) \geq \frac{v_1}{1 + \delta} \qquad \text{if } \delta \geq \hat{\delta}.$$

In (vi), S_1 must reject an offer of $v_2(1 - \delta^2/(1 + \delta)$ from B_2 in order to sustain the equilibrium. This requires that $v_2(1 - \delta^2/(1 + \delta)) \leq \delta^3 v_1/(1 + \delta)$, and explains the value of $\hat{\delta}$.

It is straightforward to check that the above strategies constitute an equilibrium. ∎

5. PRIVATE OFFERS MODEL

It is clear from the description of the equilibrium strategies in Proposition 4 that any equilibrium (for high δ) in the targetted offers model must have the feature that, off the equilibrium path, a player may reject the same offer that he would accept in equilibrium because the offer to the *other* responder has changed. This is not a particularly "intuitive" feature of equilibrium. In the equilibrium described above, it does not even result from a change in this player's expectation of future payoff. One might consider looking for equilibria in the *private offers* model, where a player's response *must* depend only on the offers he or she receives. Clearly, the results in the previous section indicate that there cannot be any *pure strategy* equilibria in the private offers model.

This raises the question of what exactly should be considered known in checking subgame perfection in the private offers model. Note that because of the imperfect information, there are no proper subgames. We borrow the concept of *public strategy* from Fudenberg *et al.* (1994). A strategy for a player i is public according to their definition if at any stage it specifies actions that depend only on publicly observable events in the past. A *public perfect equilibrium* (PPE) is defined by them to be a profile of public strategies that form a Nash equilibrium at every stage of the game.

The only public event in the private offers version of our model is the departure of a pair from the market following an agreement. We assume that contracts are announced once made. The passage of time is also public. However, we assume that this is not relevant information. (Time could have passed without any serious offers having been made.) Effectively, the mixed strategy equilibrium we characterize is a stationary equilibrium, with strategies depending only on who is in the market. However, non-stationary deviations (or for that matter non-public deviations) are not precluded. There is no advantage to a player in making such a unilateral deviation given that the others are playing stationary strategies.

In the S-game, seller S_i's strategy consists of the 4-tuple $[F_{i1}, F_{i2}, \pi_i, 1 - \pi_i]$. Here, F_{ij} is the cumulative distribution function of prices offered to B_j, $j = 1, 2$. π_i is the aggregate probability of an offer being made to B_1, while $(1 - \pi_i)$ is the probability of an offer being made to B_2.

In the B-game, buyer B_i's strategy consists of a 4-tuple $[G_{i1}, G_{i2}, \bar{\pi}_i, 1 - \bar{\pi}_i]$. These have an obvious interpretation.

Before deriving the mixed strategy equilibrium, we should state that, given the restricted dependence on history postulated above, no seller (or buyer) will make an offer that will be rejected with probability 1.[12] The only

---

[12] Proposition 6 has a case where a seller rejects any offer from B_1 if this is the only offer received, because B_2 and the other seller agree in equilibrium. However, there is a positive probability of offers from both B_1 and B_2.

136   CHATTERJEE AND DUTTA

case where delay could happen is where both sellers (buyers) make offers to the same buyer (seller), who accepts at most one of them. But, this delay is due to *coordination failure* rather than due to a player's desire to make an unacceptable offer.

LEMMA 4. *For $k = 1, 2$, F_{1k} and F_{2k} must have common support if offers to B_k are made with positive probability by both sellers.*

Proof. Suppose F_{1k} has support $[\underline{p}_1, \bar{p}_1]$, while F_{2k} has support $[\underline{p}_2, \bar{p}_2]$ with $[\underline{p}_1, \bar{p}_1] \neq [\underline{p}_2, \bar{p}_2]$.

Let $\underline{p}_1 < \underline{p}_2$, and consider two pure strategies for S_1 of offering p and p', to $\bar{B}_k$, with $p < p'$ and $p, p' \in (\underline{p}_1, \underline{p}_2)$.

Both offers must be accepted with probability 1. That is, either S_2 also offers to B_k, but p or p' is lower than any S_2 offer and is accepted, or S_2 does not make an offer to B_k, and p or p' is accepted. But then p and p' cannot both be in the support of S_1's mixed strategy since they cannot give the same payoff. The only possibility then is a mass point at $\underline{p}_1$. But, this cannot be an equilibrium strategy, since any offer in $(\underline{p}_1, \underline{p}_2)$ would do better.

Suppose $\bar{p}_1 > \bar{p}_2$. Then no S_1 offer in $(\bar{p}_2, \bar{p}_1)$ can be accepted if S_2 also makes an offer to B_k. An offer $p > \bar{p}_2$ can be accepted only if S_2 has not made an offer. But then there should be a mass point at $\bar{p}_2$. Then S_2 should deviate from $\bar{p}_2$ to just below $\bar{p}_1$ and get a higher payoff. ∎

LEMMA 5. *In the S-game, it cannot be an equilibrium for any buyer to receive two degenerate price offers.*

Proof. Suppose B_k receives offer p_1 and p_2 from S_1, S_2 with probabilities π_1, π_2, respectively. In view of Lemma 4, $p_1 = p_2$. But then either seller has an incentive to lower the price offer slightly. ∎

PROPOSITION 5. *For δ sufficiently high, there is a unique mixed strategy equilibrium in the S-game of the private offers model. The unique mixed strategy equilibrium is described by the following equations:*

$$\pi_1 F_{11}(x) = F_{21}(x) = \frac{(1 + \delta)x - v_2}{(1 + \delta)x - \delta^2 v_2} \tag{7}$$

with support $[v_2/(1 + \delta), \bar{z}]$.[13]

$$\pi_1 = \frac{\bar{z}(1 + \delta) - v_2}{\bar{z}(1 + \delta) - \delta^2 v_2} \tag{8}$$

$$\pi_2 = 1; \tag{9}$$

---

[13]The value $\bar{z}$ satisfies the condition that the buyer is better off accepting $\bar{z}$ rather than rejecting both offers and going to the B-game. The computation is shown after the next proposition.

COMPETITION FOR BARGAINING PARTNERS 137

there is a mass of S_2's probability of $1 - F_{21}(\bar{z})$ at $\bar{z}$;

$$S_1 \text{ offers } B_2 \text{ a price of } \frac{v_2}{1+\delta} \text{ with probability } (1 - \pi_1). \tag{10}$$

$$\text{The expected payoff to each seller is } \frac{v_2}{1+\delta}. \tag{11}$$

B_1's response is to accept the lowest offer less than or equal to $\bar{z}$. B_2 accepts any offer less than or equal to $v_2/(1+\delta)$. Other offers are rejected.

Proof. Let $[F_{11}, F_{12}, \pi_1, 1 - \pi_1]$, $[F_{21}, F_{22}, \pi_2, 1 - \pi_2]$ be any pair of equilibrium strategies in the S-game. Let $E_{ij}(x)$ be the expected payoff to S_i when a price of x is offered to B_j.

To calculate $E_{11}(x)$, note that $(1 - \pi_2 + \pi_2(1 - F_{21}(x))$ is the probability that B_1 will accept an offer of x from S_1, while $\pi_2 F_{21}(x)$ is the probability that such an offer will be rejected. In the latter case, S_1 obtains $\delta^2 v_2/(1+\delta)$ from the Rubinstein game with B_2.

Hence,

$$E_{11}(x) = x\big(1 - \pi_2 + \pi_2(1 - F_{21}(x))\big) + \pi_2 F_{21}(x)\frac{\delta^2 v_2}{1+\delta} = K_1. \tag{12}$$

From (12),

$$\pi_2 F_{21}(x) = \frac{(1+\delta)(x - K_1)}{(1+\delta)x - \delta^2 v_2}. \tag{13}$$

We now calculate $E_{12}(x)$.

Note that $E_{11}(x) = E_{12}(x)$ since all (pure) strategies must give S_1 the same expected payoff. So,

$$E_{12}(x) = x\big(\pi_2 + (1 - \pi_2)(1 - F_{22}(x))\big) + (1 - \pi_2)F_{22}(x)\frac{\delta^2 v_1}{1+\delta} \tag{14}$$

$$= K_1.$$

From (14), $(1 - \pi_2)F_{22}(x) = (1+\delta)(x - K_1)/((1+\delta)x - \delta^2 v_1)$. In order for F_{22} to be increasing in x, we need $K_1 > \delta^2 v_1/(1+\delta)$. So, $x > K_1 > \delta^2 v_1/(1+\delta)$. But then, for all $\delta^2 > v_2/v_1$, $v_2 - \delta^2 v_1/(1+\delta) \leq \delta v_2/(1+\delta)$. This implies that B_2 should reject any offer of x whenever B_2 receives only one offer. Hence,

$$E_{12}(x) = x(1 - \pi_2)(1 - F_{22}(x)) + \pi_2\frac{\delta^2 v_2}{1+\delta}$$

$$+ (1 - \pi_2)F_{22}(x)\frac{\delta^2 v_1}{1+\delta} = K_1,$$

or

$$F_{22}(x) = \frac{K_1 - \pi_2(\delta^2 v_2/(1+\delta)) - x(1-\pi_2)}{(1-\pi_2)(\delta^2 v_1/(1+\delta)) - x(1-\pi_2)}. \tag{15}$$

The last equation implies that either F_{22} is a decreasing function of x or $F_{22}(x) > 1$, which is impossible. Hence, $\pi_2 = 1$ if S_1 bids to B_2. Now, because S_2 only bids to B_1, S_1 only uses a pure strategy bid $v_2/(1+\delta)$ to B_2. This implies that $K_1 = v_2/(1+\delta)$. Since F_{11} and F_{21} have common support, expected payoffs to both sellers must be $v_2/(1+\delta)$.

Finally, note that by analogy, $\pi_1 F_{11}(x) = ((1+\delta)x - v_2)/((1+\delta)x - \delta^2 v_2)$. Setting $F_{11}(\bar{z}) = 1$, we get $\pi_1 = ((1+\delta)\bar{z} - v_2)/((1+\delta)\bar{z} - \delta^2 v_2)$.

The equilibrium behavior of the buyers is justified by the description of equilibrium in the B-game in the next proposition.

This completes the proof of Proposition 5. ∎

Remark 4. Note that uniqueness is encapsulated in the following elements of the proof:

(i) If both sellers offer exclusively to the same buyer, it must be to B_1 and this would entail both S_1 and S_2 having mass points at $\bar{z}$. This is ruled out in equilibrium. (A seller could achieve greater expected profits by shifting the mass downward.)

(ii) Each seller offers to a single buyer. If so, the equilibrium must be in pure strategies and this is ruled out.

(iii) Both sellers offer to both buyers with positive probability. This means $1 > \pi_i > 0$, $i = 1, 2$. This is ruled out in the proof above.

(iv) The only remaining option if an equilibrium of this type exists, is for one seller to make offers with positive probability to both buyers and the other seller to offer solely to B_1.

We now turn our attention to the buyers' game. For $i = 1, 2$, let $[G_{i1}, G_{i2}, \bar{\pi}_i, 1 - \bar{\pi}_i]$ denote a pair of equilibrium strategies of B_i. Let $R_{ij}(x)$ denote the expected payoff of B_i when x is the price offered to S_j, and let L_i be B_i's expected payoff. Then,

$$R_{11}(x) = (v_1 - x)\bar{\pi}_2 G_{21}(x)$$
$$+ \left[1 - \bar{\pi}_2 + \bar{\pi}_2(1 - G_{21}(x))\right]\frac{\delta^2 v_1}{1+\delta} = L_1. \tag{16}$$

Hence,

$$\bar{\pi}_2 G_{21}(x) = \frac{L_1 - \delta^2 v_1/(1+\delta)}{v_1 - x - \delta^2 v_1/(1+\delta)}. \tag{17}$$

COMPETITION FOR BARGAINING PARTNERS 139

Also,

$$R_{12}(x) = (v_1 - x)\left[(1 - \bar{\pi}_2)G_{22}(x)\right]$$
$$+ \left[(1 - \bar{\pi}_2)(1 - G_{22}(x)) + \bar{\pi}_2\right]\frac{\delta^2 v_1}{1 + \delta} = L_1 \tag{18}$$

Therefore,

$$(1 - \bar{\pi}_2)G_{22}(x) = \frac{L_1 - \delta^2 v_1/(1 + \delta)}{v_1 - x - \delta^2 v_1/(1 + \delta)}. \tag{19}$$

Similarly,

$$R_{21}(x) = (v_2 - x)\left[\bar{\pi}_1 G_{11}(x) + (1 - \bar{\pi}_1)\right]$$
$$+ \bar{\pi}_1(1 - G_{11}(x))\frac{\delta^2 v_2}{1 + \delta} = L_2 \tag{20}$$

Hence,

$$\bar{\pi}_1 G_{11}(x) = \frac{L_2 - \bar{\pi}_1(\delta^2 v_2/(1 + \delta)) - (1 - \bar{\pi}_1)(v_2 - x)}{v_2 - x - (\delta^2 v_2/(1 + \delta))}. \tag{21}$$

$$(1 - \bar{\pi}_1)G_{12}(x) = \frac{L_2 - (1 - \bar{\pi}_1)(\delta^2 v_2/(1 + \delta)) - \bar{\pi}_1(v_2 - x)}{v_2 - x - (\delta^2 v_2/(1 + \delta)).} \tag{22}$$

Note that B_1 has to be indifferent between offering to S_1 or S_2 when $\bar{\pi}_1$ is positive. Setting $G_{21}(\bar{x}_1) = 1 = G_{22}(\bar{x}_2)$, we get from (16) and (18),

$$L_1 = \frac{\delta^2 v_1}{1 + \delta}(1 - \bar{\pi}_2) + (v_1 - \bar{x}_1)\bar{\pi}_2 \tag{23}$$

$$= \frac{\delta^2 v_1}{1 + \delta}\bar{\pi}_2 + (v_1 - \bar{x}_2)(1 - \bar{\pi}_2). \tag{24}$$

Similarly, we can obtain

$$L_2 = \frac{\delta^2 v_2}{1 + \delta}\bar{\pi}_1 + (v_2 - \underline{x}_1)(1 - \bar{\pi}_1) \tag{25}$$

$$= \frac{\delta^2 v_2}{1 + \delta}(1 - \bar{\pi}_1) + (v_2 - \underline{x}_2)\bar{\pi}_1. \tag{26}$$

LEMMA 6. *In Eqs. (23)–(25),*

$$\underline{x}_1 = \underline{x}_2 = \frac{\delta v_2}{1 + \delta} \quad and \quad \bar{x}_1 = \bar{x}_2 = \frac{v_2}{2}\left(\frac{1 + 2\delta - \delta^2}{1 + \delta}\right).$$

Proof. Let $[\underline{y}_j, \bar{y}_j]$ be the support of B_1's offer to S_j, and let $[\underline{w}_j, \bar{w}_j]$ denote the support of B_2's offers. Suppose $\underline{y}_j, \underline{w}_j > \delta v_2/(1 + \delta)$. An argument analogous to Lemma 4 shows that $\underline{y}_j = \underline{w}_j = \underline{x}_j$. If B_2 makes an offer of $\underline{x}_j$, this wins with positive probability only if B_1 makes an offer to S_i, $i \neq j$. But, if B_2 faces no competition, the optimal offer is $\delta v_2/(1 + \delta)$. Therefore, in equilibrium, $\underline{x}_j = \delta v_2/(1 + \delta)$ for $j = 1, 2$. Therefore, $\underline{x}_1 = \underline{x}_2$.

We now show that $\bar{x}_1 = \bar{x}_2$. Suppose $\bar{x}_1 > \bar{x}_2$. Now, suppose B_2 deviates and switches some probability mass M from S_1 to S_2. Then, he gains $M(v_2 - \bar{x}_2)$. The loss is L, where

$$
\begin{aligned}
L &= \int_{\bar{x}_2}^{\bar{x}_1} \left\{ (v_2 - x)\left(\frac{1}{2} + \frac{1}{2}G_{11}(x)\right) + \frac{1}{2}\left(1 - G_{11}(x)\frac{\delta^2 v_2}{1 + \delta}\right) \right\} g_{21}(x)\, dx \\
&< \int_{\bar{x}_2}^{\bar{x}_1} (v_2 - x)g_{21}(x)\, dx \\
&< (v_2 - \bar{x}_2)M \qquad \text{where } M = G_{21}(\bar{x}_1) - G_{21}(\bar{x}_2).
\end{aligned}
$$

Therefore, $\bar{x}_1 = \bar{x}_2 = \bar{x}$. This also yields $\bar{\pi}_i = \frac{1}{2}$ for $i = 1, 2$.

To calculate $\bar{x}$, substitute $\underline{x} = \delta v_2/(1 + \delta)$ in (25) to obtain $L_2 = ((1 + \delta^2)/(2(1 + \delta)))v_2$. We substitute this into (21). Also, use $\bar{\pi}_1 = \frac{1}{2}$ and calculate the value $\bar{x}$ such that $G_{11}(\bar{x}) = 1$. This yields

$$
\bar{x} = \frac{v_2}{2}\left(\frac{1 + 2\delta - \delta^2}{1 + \delta}\right).
$$

PROPOSITION 6. *For δ sufficiently high, there is a unique mixed strategy equilibrium in the B-game of the private offers model. The unique mixed strategy equilibrium is described by the following:*

$$
G_{1i}(x) = \frac{2L_2 - \delta^2 v_2/(1 + \delta) - (v_2 - x)}{v_2 - x - \delta^2 v_2/(1 + \delta)} \tag{27}
$$

$$
G_{2i}(x) = \frac{2(L_1 - \delta^2 v_1/(1 + \delta))}{v_1 - x - \delta^2 v_1/(1 + \delta)} \tag{28}
$$

with common supports

$$
\left[\frac{\delta v_2}{1 + \delta}, \frac{(1 + 2\delta - \delta^2)}{2(1 + \delta)}v_2\right] \tag{29}
$$

$$
\bar{\pi}_1 = \bar{\pi}_2 = \tfrac{1}{2}, \tag{30}
$$

G_{2i} *has a probability mass at* $\underline{x} = \delta v_2/(1 + \delta)$,

$$
L_1 = \frac{1}{2}\left[v_1\left(\frac{1 + \delta + \delta^2}{1 + \delta}\right) - \frac{v_2}{2}\frac{(1 + 2\delta - \delta^2)}{1 + \delta}\right] \tag{31}
$$

$$
L_2 = \frac{1 + \delta^2}{2(1 + \delta)}v_2. \tag{32}
$$

COMPETITION FOR BARGAINING PARTNERS   141

Seller S_i rejects any offer less than $\delta v_1/(1+\delta)$ from B_1, if this is the only offer received. Offers from B_2 or offers from both buyers are rejected if less than $\delta v_2/(1+\delta)$.

Proof. Lemma 6 has established (29) and (30). Substitution of $\bar{\pi}_2 = \frac{1}{2}$ in (17) and (19) gives (28), while $\bar{\pi}_1 = \frac{1}{2}$ together with (21) and (22) establishes (27). Substitution of $\underline{x}$ and $\bar{x}$ from (29) into (23) and (25) yields (31) and (32).

Optimality of the seller responses can be checked as follows. If S_i receives an offer $p < \delta v_1/(1+\delta)$ from B_1 alone, rejection generates the Rubinstein subgame with B_1 (B_2 and S_j then have agreed in equilibrium), in which the seller's payoff is $v_1/(1+\delta)$, but one period later. Therefore, rejection is optimal.

If S_i receives offers from both buyers, a rejection generates the S-game in which the seller's expected payoff is $v_2/(1+\delta)$, again one period later. A rejection of an offer from B_2 alone also gives the S-game as above. Therefore, S_i should accept any offer greater than or equal to $\delta v_2/(1+\delta)$. In order to show uniqueness, we note the following:

(i) No pure strategy equilibrium exists.

(ii) If B_i were to make offers only to S_i, $i = 1, 2$, the offers would have to be pure strategies, and hence would not be equilibria under (i).

(iii) Therefore, the only other possibility is if B_1 makes offers only to S_1, say, and B_2 makes offers to both sellers. In this case, the upper bound of the support for the mixed strategy to S_1 would be $\delta v_2/(1+\delta)$. The mixed strategies then have mass points at $\underline{x}$. (The details of the calculations are omitted.) But then either B_i could deviate and shift his or her mass point a small amount upward, increasing expected payoff. So, this cannot be an equilibrium.

The case where B_1 offers to both is analogous. In order to complete the check that the strategies described above constitute an equilibrium, we also need to verify that the sellers in the S-game and buyers in B-game do not gain by making rejected offers. The expected seller payoff in the S-game is $v_2/(1+\delta)$, while the *maximum* seller payoff in the B-game is $((1 + 2\delta - \delta^2)/2(1+\delta))v_2$. Hence, sellers are better off in the S-game. Similarly, both buyers are better off in the B-game.

Note that $\bar{z}$, the maximum price offered by sellers to B_1, is irrelevant to the sellers' expected payoffs so long as $\bar{z} > v_2/(1+\delta)$. But, $\bar{z}$ must satisfy the restriction

$$v_1 - \bar{z} \geq \max\left(\delta L_1, \frac{\delta v_1}{1+\delta}\right). \tag{33}$$

Relation (33) ensures that B_1 is willing to accept $\bar{z}$ in the S-game. For δ sufficiently large, $L_1 \geq v_1/(1+\delta)$, so that $\bar{z} = v_1 - \delta L_1$.

This completes the proof of Proposition 6. ∎

Of course, the mixed strategy equilibrium involves coordination problems, usual in simultaneous move models, that generate a probability of delay. However, we find the mixed strategy equilibrium intuitively more plausible than the pure strategy equilibria in the previous models.

The intuition supporting this equilibrium is the following. Both sellers want to be matched with B_1; this is the competitive or auction-like element. However, the presence of B_2 limits the extent to which the sellers need to compete. One of them can switch to B_2 and ensure the bargaining payoff. When the buyers make offers, B_1 wants to avoid the bargaining game and makes low offers to the sellers, who accept if both B_1 and B_2 make offers to the same seller, and reject B_1 offers otherwise. The sellers accept low offers because they want to avoid the auction-like setting of the S-game. However, each of them wants to bargain with B_1 if B_2 is out of the market. Thus, auctions and the bargaining aspects mesh together—the bargaining sets an endogeneous reservation price in the S-game auction, while the desire to stay out of the auction drives seller acceptance of low bids in the B-game.

When buyers make offers, prices essentially converge to $v_2/2$, the lower of the two "Rubinstein" prices. When sellers make offers, sellers' *expected* payoffs also equal $v_2/2$. However, uniform pricing does not prevail since the support of the (equilibrium) mixed strategy remains nondegenerate even for "high" δ. Also, the allocation is inefficient because of the positive probability of delay.

6. CONCLUSION

In this paper, we have considered a small market with features of both bargaining and auctions under complete information. Table II summarizes the limiting values (as $\delta \rightarrow 1$) of prices and payoffs for the three trading processes. (For the non-degenerate mixed strategies, we have just indicated the supports.)

The following features are notable:

1. In several cases, the results can be interpreted as an auction of bargaining rights with $v_2/2$ rather than v_2 being the auction price. This is true when buyers make offers in the public offers model as well as in the B-game of the targeted offers model. Moreover, even in the mixed strategy equilibria of the private offers model, the *expected* payoffs of S-game sellers equals $v_2/2$, and the equilibrium price in the B-game goes to $v_2/2$.

2. In the public offers model, the equilibrium when buyers make offers involves both sellers accepting the same buyer offer, thus generating

TABLE II
Outcomes of Equilibrium Play, $\delta \to 1$

I. Public offers model (pure strategy equilibrium)
S-game
    Both pairs matched at price $p^* = v_2/2$.
B-game
    The prices, though different, go to $v_2/2$.
II. Targetted offers model (pure strategy equilibrium)
S-game
    (S_2, B_2) agree in first period at price $v_2/2$.
    (B_1, S_1) agree in next period at price $v_1/2$.
B-game
    (S_2, B_1) agree in first period, (S_1, B_2) agree in second period.
    In both cases, the price is $v_2/2$.
III. Private offers model (mixed strategy equilibrium)
S-game
    S_1 and S_2 make offers to B_1 with support $[v_2/2, (v_1 + v_2)/4]$;
    one seller makes offers to B_2 of $v_2/2$.
B-game
    Both buyers make offers to both sellers with support tending to $[v_2/2, v_2/2]$.

---

delay. Also, the pure strategy equilibrium in the S-game, the only one in this paper with immediate acceptance of a uniform price, is not *robust* since even a slight difference in sellers' reservation values will destroy the equilibrium.

3. In the targetted offers model, more than one price is charged in equilibrium when sellers make offers. Also, all pure strategy equilibria in the targetted offers model exhibit delay since both pairs cannot be matched in the initial period.

4. The results, therefore, show that neither *uniform pricing* nor *efficiency* are guaranteed in decentralized trading in thin markets.

APPENDIX

Proof of Lemma 0. Note that since $(b_1, b_2, s_1, s_2) \in Q$, both S_1 and S_2 must reject their respective offers following the deviation of B_i. Otherwise, (b_1, b_2, s_1, s_2) cannot be an equilibrium of the B-game. So, there exist equilibrium seller payoffs k periods later "sustaining" the rejection decision, i.e., there exist equilibrium payoffs (s'_1, s'_2) such that $(s_1, s_2) \le \delta^k(s'_1, s'_2)$.

If (s'_1, s'_2) are S-equilibrium payoffs, then there is nothing to prove. So, let $Q' \subseteq Q$ be the set of payoff vectors for which there exist no such sustaining S-equilibria. Let s_1^* be the supremal seller payoff in Q', and let (s_1^*, s_2^*) be the equilibrium payoff associated with this. Then there exists a B-game equilibrium with seller payoffs (s_1', s_2') such that $(s_1^*, s_2^*) \leq \delta(s_1', s_2')$. So, $(s_1', s_2') \notin Q'$ by the definition of s_1^*.

Now, for $(s_1, s_2) \in Q \backslash Q'$, $(s_1, s_2) \leq \delta(m_1, m_2)$, where (m_1, m_2) are S-game equilibrium payoffs. But, by assumption, there is no S-game equilibrium payoff sustaining (s_1^*, s_2^*), and hence (s_1', s_2'). Therefore, $(s_1', s_2') \notin Q$.

So, (s_1', s_2') must be obtained from a B-game equilibrium which is not in Q. Therefore, any profitable deviation from B_i will imply that *one* seller rejects her offer. Hence, either $s_i' = \delta v_i/(1 + \delta)$ or $s_j' = \delta v_j/(1 + \delta)$. But, routine checking shows that these are the lowest seller payoffs in equilibrium, and so we cannot have $(s_1^*, s_2^*) \leq \delta^2(s_1', s_2')$. ∎

Proof of Lemma 3. We remind the reader that $\underline{\delta}$ satisfies (4). From Table I and Lemma 0, we have

$$\bar{s} \leq \max\left(\frac{v_1}{1 + \delta}, p_1^s, p_2^s\right). \tag{A.1}$$

But, for $i = 1, 2$, $p_i^s \leq v_i - \delta \underline{b}_i$, where $\underline{b}_i$ is the minimum payoff of B_i in the B-game. Hence,

$$\bar{s} \leq \max\left(\frac{v_1}{1 + \delta}, v_1 - \delta \underline{b}_1, v_2 - \delta \underline{b}_2\right). \tag{A.2}$$

We consider only the possibilities that the maximum in (A.2) is attained at $v_i - \delta \underline{b}_i$.

Suppose $v_1 - \delta \underline{b}_1 \geq v_2 - \delta \underline{b}_2$. Let $\underline{b}_1$ be obtained from either Case 4 or Case 5. Then,

$$\bar{s} \leq v_1 - \delta \underline{b}_1 \leq v_1 - \delta(v_1 - \delta \bar{s}). \tag{A.3}$$

The second inequality in (A.3) follows because no seller can reject an offer higher than $\delta \bar{s}$ from B_1. But (A.3) yields $\bar{s} \leq v_1/(1 + \delta)$. The other possibility is that $\underline{b}_1$ is obtained from Case 6, so that $\underline{b}_1 = \delta^2 v_1/(1 + \delta)$. Hence, $\bar{s} \leq v_1(1 - \delta^3/(1 + \delta))$. But then $\delta \bar{s} < v_2(1 - \delta^2/(1 + \delta))$. So, B_1 can make a price offer arbitrarily close to $v_1(1 - \delta^2/(1 + \delta))$, and this cannot be rejected by S_1. This would rule out Case 6 as an equilibrium.

Suppose $v_2 - \delta \underline{b}_2 > v_1 - \delta \underline{b}_1$. If $\underline{b}_2$ is obtained from Case 4 or Case 6, then in exactly the same way as above, it will follow that $\bar{s} \leq v_2/(1 + \delta)$. If $\underline{b}_2$ is obtained from Case 5, then $\bar{s} \leq v_2(1 - \delta^3/(1 + \delta)) \leq v_1/(1 + \delta)$ since $\delta \geq \underline{\delta}$.

This completes the proof of Lemma 3. ∎

REFERENCES

Bennett, E., and van Damme, E. (1991). "Demand Commitment Bargaining—The Case of Apex Games," in *Game Equilibrium Models III* (R. Selten, Ed.). Berlin/New York: Springer-Verlag.

Binmore, K. (1985). "Bargaining and Coalitions," in *Game Theoretic Models of Bargaining* (A. E. Roth, Ed.). Cambridge, UK: Cambridge Univ. Press.

Chatterjee, K., Dutta, B., Ray, D., and Sengupta, K. (1993). "A Non-Cooperative Theory of Coalitional Bargaining," *Rev. Econ. Studies* **60**, 463–477.

Fudenburg, D., Levine, D., and Maskin, E. (1994). "The Folk Theorem with Imperfect Public Information," *Econometrica* **62**, 997–1039.

Gale, D., (1986). "Bargaining and Competition Part I: Characterization," *Econometrica* **54**, 785–806.

Hendon, E., and Tranaes, T. (1991). "Sequential Bargaining in a Market with One Seller and Two Different Buyers," *Games Econ. Behav.*

Moldovanu, B., and Winter, E. (1992). "Order Independent Equilibria," mimeo. University of Bonn and Hebrew University of Jerusalem.

Osborne, M., and Rubinstein, A. (1990). *Bargaining and Markets*. San Diego: Academic Press.

Perry, M., and Reny, P. (1994). "A Non-Cooperative View of Coalition Formation and the Core," *Econometrica* **62**, 795–817.

Roth, A. E., and Sotomayor, M. A. O. (1990). *Two-Sided Matching: A Study in Game-Theoretic Modelling and Analysis*. Cambridge, UK: Cambridge Univ. Press.

Rubinstein, A. (1982). "Perfect Equilibrium in a Bargaining Model," *Econometrica* **50**, 97–109.

Rubinstein, A., and Wolinsky, A. (1985). "Equilibrium in a Market with Sequential Bargaining," *Econometrica* **53**, 1133–1150.

Rubinstein, A. and Wolinsky, A. (1990). "Decentralised Trading, Strategic Behavior and the Walrasian Outcome," *Rev. Econ. Studies* **57**.

Shaked, A. (1994). "Opting Out: Bazaars vs. Hi Tech," *Investigaciones Econ.* **18**(3), 421–432.

Shubik, M. (1982). *Game Theory for the Social Sciences*. Cambridge, MA: MIT Press.

Bargaining, Competition and Efficient Investment*

Kalyan Chatterjee**

Department of Economics,
The Pennsylvania State University,
University Park, Pa. 16802, USA

Y. Stephen Chiu

School of Economics and Finance
University of Hong Kong,
Hong Kong

This paper explores the interplay between choice of investment type (specific vs. general), bargaining extensive form and endogenous outside options in the framework of incomplete contracts introduced formally in the work of Grossman, Hart and Moore. We find that the bargaining procedure chosen has significant implications for choice of investment and for the usefulness of the assignment of property rights in enhancing efficiency. Somewhat paradoxically an "auction-like" procedure might need the correct assignment of property rights for a more efficient solution while a sequential offers procedure might do as well as the best assignment of property rights.

1 Introduction

1.1 Main features

The aim of this paper is to investigate investment choice by agents in a "thin" market where transactions take place by bilateral bargaining. We study the simplest non-trivial model of this type with two buyers and one seller. Our model has the following features.

1. Both seller and buyers potentially invest. The seller chooses both the type of the investment and the level. The type of the investment varies continuously from 0 (investment specific to the first buyer's product) to 1 (investment specific to the second buyer's product) with a value of $\frac{1}{2}$ being the most "generalist" type.

---

*Based on a 2006 draft.
**We thank Hongbin Cai, George Mailath, Abhinay Muthoo and Andrew Postlewaite for useful comments.

Investment specific to one buyer creates the most value in a transaction with that buyer per unit of investment chosen and the least value in a transaction with the other buyer. We assume the seller's investment is "more important" than the buyers' in a sense to be explained later.

2. The environment is one of incomplete contracting. Neither investment level or type or value can be contracted on. Each agent owns an asset in the benchmark model and the investment is in learning how to use the asset for a specific purpose, either for one specific buyer or for something generalist. (The investment is "inalienable").

3. The "outside option" of the seller in a negotiation with one buyer is the value of an agreement with the other buyer. (The alternative to a negotiated agreement for a buyer is an inside option, that is he produces the item himself.) Since the bargaining institution affects not only the payoff within a negotiation but also the outside option, the nature of the investment chosen depends crucially on it. We consider two extensive forms, which we think of as corresponding to different kinds of market (as in Shaked (1994), though our markets are a bit different from his). In a Bertrand-type "bazaar", the short side of the market (here the seller) obtains all the surplus while the long side gets nothing. This is also in the core of the bargaining game. Given the assumptions made, this has the consequence of providing appropriate incentives for the seller to choose the efficient *level* of investment. However, this need not lead to efficiency. In the second market, probably more relevant to modern industry, a transaction involves a single buyer and seller exchanging offers in any one period (like a merger negotiation), and the short side does not get the entire surplus. Paradoxically, this could turn out to be more efficient in terms of choice of investment.

4. The most important feature of our model is the interplay between investment choice and bargaining. Bargaining is sometimes modelled in literature of this kind as an arbitrary division of the total surplus in negotiation and renegotiation and the consequences for investment choice are investigated. In this paper, the extensive form is fixed but the share of surplus is endogenously determined by the investment itself. It is not possible here to consider these two features separately.

5. The efficient assignment of property rights turns out to depend on the type of market we are considering. If it is a "bazaar", assigning property rights to a buyer for the asset of the seller in addition to her own asset might improve efficiency by ensuring the right choice of type of investment. For the other case, it makes no difference.

In order to fix our minds on the nature of the phenomena being studied, consider a firm of marketing consultants who can invest in specialised knowledge about one industry (say biotechnology) or invest in skills that are as relevant for diet soft drinks as for the biotechnology industry. The type of investment and the amount of investment both matter here. A similar idea in the academic labour market is mentioned in Chatterjee and Marshall (2003); an academic who chooses to invest in her own speciality will become more valuable in that speciality but will lose a potential job opportunity in, say, a consulting firm.

1.2 Related literature

The paper most closely related to this is our own earlier work on investment and competition (Chatterjee and Chiu, 2000). That paper essentially dealt with the choice of type of investment and considered the "bazaar-like" extensive form leading to a core allocation, showing that competition could increase inefficiency. In that paper, only one side of the market had to choose investment. Further, the interplay between level and type of investment and the bargaining procedure was not considered in our earlier paper, though the single bargaining procedure considered had m sellers and n buyers, with $n > m \geq 1$. Felli and Roberts (2002) and Cole, Mailath and Postlewaite (2001a,b) discuss a similar problem though the latter set of authors analyses the market part by using co-operative game theory.[1] Felli and Roberts do use a Bertrand-type mechanism for choosing allocations and prices but again only one side of the market invests. Both groups of authors obtain efficiency in any mechanism where the bargaining process leads to an allocation in the core and neither considers the nature of the investment, only its level. De Meza and Lockwood (1998b) analyse a somewhat different search and bargaining model in which sellers do not know a priori which buyer they will meet in the transaction phase. They use this as an explanation why complete contracts cannot be written before players choose investment levels.

The paper by Cai (2003) addresses the issue of specific investment, though in a somewhat different way from this paper. His model has two kinds of investment, specific, in the relationship and general, in the outside option. Thus he chooses a more "reduced form" approach than we do in this paper; his paper has no explicit bargaining and the outside option and its change with general investment are exogenously specified. In our model, there is only one investment for the seller, who must, however, choose how specific she wants it to be. The seller therefore chooses her market power but is constrained by the nature of the market (extensive form) in determining the extent to which this power translates into higher payoffs. In the first extensive form we consider, joint ownership would not in fact be an optimal assignment of property rights; such rights should be assigned to the buyer. (The conclusion therefore also differs from Cai's paper.)

While these papers are the most relevant for understanding competition and specificity, there is a long list of papers on incomplete contracts and property rights beginning with the seminal work of Grossman and Hart (1986) and Hart and Moore (1990). Gans (2003) has an interesting variation in which property rights are not assigned but assets are sold by auction. Chiu (1998), de Meza and Lockwood (1998a) and Rajan and Zingales (1998) explore how the optimal assignment of property

---

[1] As they have pointed out, core allocations can also be obtained through several non-cooperative procedures and the absence of discounting in their paper is a more substantial difference.

rights in Hart and Moore can be reversed by considering outside options as they appear in the strategic bargaining literature starting from Rubinstein (1982). The relevant bargaining literature on outside and inside options is ably surveyed by Muthoo (1999).

2 The Model

There are three agents, one seller (S) and two buyers $(B_1$ and $B_2)$. The seller can make one unit of the good or service; each buyer has a demand for at most one unit. The maximum price a buyer i is willing to pay depends on the value v_i, which can be enhanced by seller investment in human capital. The buyer can also produce the item himself in which case the value to him will be $\underline{v}(=0$, for convenience).

We consider two extensive form games. In both of these the seller has to choose γ, the investment in knowledge, and the cost of the investment is given by $c(\gamma)$, which is increasing, strictly convex, and differentiable everywhere. Provided that the investment γ is fully specific to buyer B_i, the value to the buyer B_i is given by

$$v_i = v_{0i} + \alpha\gamma, \tag{1}$$

where α is a coefficient and v_{0i} is a component resulting from investment by buyer B_i.

In case the investment is not fully specific to the buyer, the value is reduced. This is captured by the seller's choice of x. The specification of the investment type is more fully described in Section 4.1.

Both seller and buyers' investment decisions are made at time 0. Bargaining occurs at time 1.

The extensive forms of the bargaining differ in what happens subsequent to the investment being made. The investment, in common with the usual practice in this literature, is observable but contracts cannot be written ex ante in which the terms differ for different values of investment.

The following extensive forms are considered.

I. Auction-like procedure The buyers, B_1 and B_2 make price offers to the seller, who accepts at most one.[2] If an offer is accepted the game ends; if not, the seller makes a counter offer in the next period to one of the buyers who then accepts or rejects and so on. Each time there is a rejection, payoffs are discounted as is usual in bargaining theory.

II. Sequential procedure The bargaining in the second extensive form is sequential; in the first period a randomly chosen buyer makes an initial offer to the seller who either accepts the offer or rejects it and proceeds to the next period (an even period) to either make a counter-offer to the same buyer or switch to

---

[2] Here the distinction made in Chatterjee and Dutta (1998) about random matching and strategic choice of partners does not have any relevance, since the seller can choose one specific buyer when she makes a counter-offer. The key here is the auction-like procedure of competition among buyers.

the other buyer and make an offer. Again payoffs obtained in periods after a rejection are appropriately discounted.

Any acceptance ends the game. All agents have the same discount factor. Note that while there is discounting during bargaining, there is not between time 0 and time 1, at which the bargaining starts.

The two different bargaining procedures correspond to different institutional settings. The first one is most "market-like" or "auction-like" though, unlike in an auction, the game could extend for longer than one period. The interaction between bargaining procedure and the investment decision is at the heart of this paper. The bargaining procedures itself have been studied (without investment choice), for example in Osborne and Rubinstein (1990) and Chatterjee and Dutta (1998).

In the example we discussed briefly in the last section, the second extensive form appears more natural, since a buyer seeking to hire a firm of marketing consultants would presumably want to meet them separately first and then make an offer, rather than calling on the consultants to call out a price for their services. This implicit choice of extensive form could be due to uncertainty about the quality of the consultants or about hesitation in sharing proprietary information, though we do not model these factors explicitly in this paper.

2.1 Payoffs and seller/buyer investment

Suppose seller S invests γ, obtains a price p at time t. Then her payoff is

$$\delta^{t-1}p - c(\gamma),$$

where δ is the common discount factor and $c(\gamma)$, as mentioned earlier, is the cost of investment.

Buyer i has a payoff of $\delta^{t-1}(v_i - p)$ if he purchases from the seller and pays a price p at time t, and obtains $\delta^{t-1}\underline{v}$ if he chooses to opt out and produce himself at time t. In Section 4.2, we consider buyer investment in determining v_{0i}. There $v_{0i} = \beta a_i$, where a_i is the investment made by B_i, and the cost of investment is again $c(a_i)$.

3 Equilibrium of Extensive Form 1 (The "Auction-Like" Mechanism)

We consider pure strategy subgame perfect equilibria of this game. Therefore, we first obtain the equilibria in the bargaining subgame. Recall this begins with the buyers making simultaneous offers to the seller who can choose to accept or reject.[3]

---

[3] A question that has been asked concerns the order of moves-would it matter if the seller moved first in making offers. Such a game would be identical with our game if the initial seller offer were to be rejected. The first period would therefore have a unique continuation payoff in the event of a rejection and the seller's payoff would be different only to the extent that discounting makes it preferable to be a proposer rather than a responder.

The values v_1 and v_2 are, of course, endogenously determined by seller investment. We therefore have to consider all possible combinations of these quantities.

Case (i) The outside option $\underline{v} > v_1, v_2$. In this (trivial) case, there is no trade and B_1, B_2 take their outside options. We rule this out by assuming $v_i > 0 = \underline{v}$

Case (ii) The maximum price B_2 can offer S, v_2, is lower than the price S would obtain by bilateral bargaining with B_1 alone (assuming for the moment that $v_1 > v_2 \geq \underline{v} = 0$). That is $v_2 \leq \frac{\delta v_1}{1+\delta}$. In this case, B_2 is irrelevant in the post-investment matching and bargaining and S and B_1 play a bargaining game.

Case (iii) Keeping the assumption that $v_1 \geq v_2$, assume now that $v_2 > \frac{\delta v_1}{1+\delta}$.

Lemma 1. *In case* (iii), *the following strategies constitute a subgame perfect equilibrium in the bargaining subgame.*

1. *B_1 and B_2 make offers of v_2 whenever it is the buyers' turn to make an offer and both buyers are in the game (neither has exercised his outside option).*
2. *Suppose both buyers are present. Whenever one or both of the offers from them are at least v_2, the seller accepts the higher one and the one from B_1 in case of a tie. The seller never accepts any offer below v_2.*
3. *S asks B_1 for a price p such that $v_1 - p = \delta(v_1 - v_2)$ and B_1 accepts. B_1 accepts any price less than it and rejects any higher price.*
4. *If only $B_i, i = 1, 2$, is present the strategies followed are identical with the Rubinstein alternating-offers bargaining game.*

Proof. It is clear that the strategies above constitute a subgame perfect equilibrium in the bargaining subgame. $\square$

Remark 1. It is also clear that the outcome is also the unique subgame perfect equilibrium outcome. If there is only one buyer (the other opts out), the equilibrium payoff is uniquely given by Rubinstein's result. Suppose therefore that both the buyers are present. Any $p < v_2$ being accepted by the seller cannot be an equilibrium outcome because buyers will bid it up. Suppose there is an equilibrium in which the maximal bargaining payoff of the seller in any equilibrium is $M_S > v_1 - \delta(v_1 - v_2) > \frac{v_1}{1+\delta}$. B_1 should reject and offer δM_S; and this will be accepted and give B_1 a higher payoff since $v_1(1 - \delta) - M_S(1 - \delta^2) > 0$. A higher payoff for a buyer can be ruled out in similar fashion.

4 Investment in the "Auction-like" Mechanism

4.1 The seller investment decision

We now consider the first stage of the game where players invest to increase the surplus before the bargaining takes place.

In this subsection only the seller invests. In the next subsection we shall consider investment by the buyers.

The seller can choose a type of investment, $x \in [0, 1]$, as well as a level of investment γ. The value of the item to B_1 will then be

$$v_{01} + g(1 - x) \bullet \alpha\gamma$$

and to B_2

$$v_{02} + g(x) \bullet \alpha\gamma.$$

Here $g(\bullet) : [0, 1] \longrightarrow [0, 1]$ is an "effectiveness" function. We assume $g(0) = 0$, $g(1) = 1$, g strictly concave, monotonic and twice differentiable everywhere. The strict concavity assumption is there to ensure that lack of specificity of investment does not cause "too much" loss in output, since $g(\frac{1}{2}) > \frac{1}{2}$.

Thus, if $x = 0$ or 1, the seller specializes his knowledge acquisition to the needs of a specific buyer, while for any $x \in (0, 1)$, the investment is of value to both buyers. The cost of investment γ is given by a strictly convex function $c(\gamma)$, while the choice of x is costless.

We now characterize the optimal choice of investment of the seller.

Proposition 1. *Suppose v_{01} and v_{02} are known and are sufficiently close.*[4] *Then Player S will choose x and γ in such a way that $v_1 = v_2$. In particular, when $v_{01} = v_{02}$, she will choose $x = \frac{1}{2}$ and $\gamma = \gamma^*$ where*

$$c'(\gamma^*) = g\left(\frac{1}{2}\right) \bullet \alpha. \tag{2}$$

Proof. Suppose on the contrary $v_1 > v_2$. There are two cases.

Case (i) $v_1 > v_2 > \frac{\delta}{1+\delta}v_1$. Then the price paid by B_1 is v_2. By increasing x, v_2 will increase for the same value of γ, and therefore the original value of x could not have been optimal.

Case (ii) $\frac{\delta}{1+\delta}v_1 \geq v_2$. Then, from the discussion preceding Lemma 1, the price paid by B_1 is $\frac{\delta}{1+\delta}v_1$. To maximize this value, since B_2 here becomes irrelevant to the payoff for S, the seller's optimal choices of x and γ must be such that is $x = 0$ and $c'(\gamma) = \frac{\delta}{1+\delta} \bullet \alpha$. S's payoff is then

$$\frac{\delta}{1+\delta}[v_{01} + g(1) \bullet \alpha\gamma] - c(\gamma).$$

But for the same value of γ, suppose x is chosen so as to make $v_2' = \frac{v_1}{2}$,[5] that is

$$v_{02} + g(x) \bullet \alpha\gamma = \frac{1}{2}[v_{01} + \alpha\gamma]. \tag{3}$$

---

[4]See section 4.2 and the next footnote.
[5]This is possible if the difference between v_{01} and v_{02} is not too great, as is implied by the condition that the buyer investment is less important than the seller investment. If βa^* is the highest possible value of buyer investment (see Section 4.2) then this is possible if there exists a x such that $\frac{1}{2}[\beta a^* + \alpha\gamma] = g(x)\alpha\gamma$. This is certainly satisfied if $\beta a^* < \alpha\gamma$, where $\frac{\delta}{1+\delta}\alpha = c'(\gamma)$, since some x less than unity will have the desired effect.

Now $v_1' = v_{01} + g(1-x)\alpha\gamma$, and $v_1' - v_2' = v_{01} - v_{02} + g(1-x)\alpha\gamma - g(x)\bullet\alpha\gamma$. Using (3), we have

$$v_1' - v_2' = v_{02} + g(x)\bullet\alpha\gamma + g(1-x)\alpha\gamma - g(1)\alpha\gamma,$$

$$= v_{02} + \alpha\gamma[g(x) + g(1-x) - g(1)] > 0.^6$$

Therefore, for the same value of γ, the seller could obtain a payoff of $v_2' > \frac{\delta}{1+\delta}v_1$. This is contradictory to the claim that $\frac{\delta}{1+\delta}v_1 \geq v_2$.

The cases where $v_2 > v_1$ are symmetric to the ones considered here and similar reasoning leads to the same conclusions as here. This completes the proof that $v_1 = v_2$. The specific results for the case of $v_{01} = v_{02}$ are straightforward and omitted.   □

4.1.1 *Comparison with the single buyer case*

The presence of a second buyer has two opposing effects on seller investment, compared to the case of a single buyer. With a single buyer, S invests γ' such that

$$c'(\gamma') = \frac{\delta}{1+\delta}\bullet\alpha,$$

and $x = 0$ or 1.

With two (symmetric) buyers, $x = \frac{1}{2}$ and

$$c'(\gamma^*) = g\left(\frac{1}{2}\right)\bullet\alpha,$$

with

$$g\left(\frac{1}{2}\right) > \frac{1}{2} > \frac{\delta}{1+\delta}.$$

Only one buyer can be served by the seller so the social surplus due to the investment is

$$v_0 + \alpha\gamma' - c(\gamma')$$

for the first case and

$$v_0 + g\left(\frac{1}{2}\right)\bullet\alpha\gamma^* - c(\gamma^*)$$

in the second case (to allow for proper comparison, here we assume $v_{01} = v_{02} = v_0$).

The two effects are that: (i) $g(\frac{1}{2}) < 1$ and therefore the social surplus is smaller when B_1 and B_2 are both present, for the same fixed amount of investment, and (ii) $\gamma^* > \gamma'$ because the seller obtains all the benefit from increasing investment with two buyers present.

In general, it is not possible to specify which effect will dominate, since this depends on how close $g(\frac{1}{2})$ is to $\frac{1}{2}$ and γ^* to γ'.

---

[6]This depends on the buyers making offers in the first period. If the seller makes offers, the result holds for sufficiently high δ.

Example 1: Suppose $c(\gamma) = \frac{1}{2}\gamma^2$ and $g(\frac{1}{2}) = \frac{1}{2} + \varepsilon$, and $\delta \approx 1$.
Then

$$c'(\gamma^*) = \gamma^* = g\left(\frac{1}{2}\right) \bullet \alpha = \alpha\left(\frac{1}{2} + \varepsilon\right).$$

$$c'(\gamma') = \gamma' = \frac{1}{2} \bullet \alpha$$

(i) Total surplus with B_1 alone

$$= v_0 + \alpha \bullet \frac{1}{2}\alpha - \frac{1}{2}\frac{1}{4} \bullet \alpha^2.$$

$$= v_0 + \frac{3}{8}\alpha^2.$$

(ii) Total surplus with B_1 and B_2

$$= v_0 + \left(\frac{1}{2} + \varepsilon\right) \bullet \alpha \bullet \alpha\left(\frac{1}{2} + \varepsilon\right) - \frac{1}{2}\left(\left(\frac{1}{2} + \varepsilon\right)\alpha\right)^2$$

$$= v_0 + \alpha^2\left(\frac{1}{2} + \varepsilon\right)^2\left(1 - \frac{1}{2}\right)$$

$$= v_0 + \alpha^2 \bullet \frac{1}{2}\left(\frac{1}{2} + \varepsilon\right)^2$$

When ε is close to 0, this is clearly less than the surplus with B_1 alone. If $\varepsilon = \frac{1}{2}$, its largest possible value, then the total surplus in (ii) is higher.

Example 2: Here the $c(\bullet)$ does not satisfy strict convexity since we assume it to be linear ($c(\gamma) = c\gamma$) and suppose that γ must be between 0 and 1 (both inclusive).

Again with δ close to 1, the payoff to the seller from case (i) is

$$\frac{1}{2}[v_0 + \alpha\gamma] - c\gamma.$$

Obviously, $\gamma' = 1$ if $\frac{1}{2}\alpha > c$ and 0 otherwise.
Similarly, from case (ii), $\gamma^* = 1$ if $(g(\frac{1}{2}) \bullet \alpha - c) > 0$ and 0 otherwise.
The total surplus will be lower under (ii), except if $\gamma^* = 1$ and $\gamma' = 0$, in which case the addition of the second buyer has a beneficial effect.

There is, therefore, a non-trivial subset of parameter values for which the two buyer/one seller auction-like mechanism reduces the total surplus from the transaction, in the absence of verifiable and enforceable contracts.

4.2 Buyer investment with the auction-like mechanism

Suppose now that the quantity v_{0i} in the expression for the value produced by a seller-buyer transaction is in fact dependent on investment by the buyer concerned. We assume that this investment is done at time 0 at the same that seller's choices of γ and x are made, before the bargaining begins. Let $v_{0i} = \beta a_i$, where a_i is the amount of investment chosen by buyer $B_i, i = 1, 2$. Suppose the cost of the investment to B_i is given by a strictly convex function again, namely $c(a_i), i = 1, 2$. The buyers are identical in this respect as well, that they have the same cost functions.

To recall, we are going to assume here that the buyer investment, which affects v_{0i}, is "less important" than the seller investment in the following sense. First of all, define a_i^* so that

$$\beta = c'(a_i^*). \tag{4}$$

This is the largest value of investment that a rational buyer can possibly choose in equilibrium. We impose the following restriction on a_i^*.

If

$$v = \frac{\delta}{(1+\delta)} \alpha \gamma^*,$$

where

$$c'(\gamma^*) = \frac{\delta}{1+\delta} \bullet \alpha, \tag{5}$$

then

$$v > \beta a_i^* \quad \text{for } i = 1, 2.$$

Then for the maximum possible value of βa_i^*, the seller can choose a x to get a higher payoff than she would with investment specific to buyer i. Note that $g(\frac{1}{2}) > \frac{1}{2} > \frac{\delta}{1+\delta}$.

The above condition guarantees that the seller will prefer to have the two buyers compete away their surplus rather than obtaining half of the surplus with the buyer who has invested more.

The above restriction is a simplifying assumption; relaxing it would not change the basic qualitative conclusions of this paper, though it could give rise to asymmetric equilibria in the auction-like mechanism.[7]

Proposition 2. *Suppose that S chooses an investment level γ^* (as defined in (5)) and sets $x = \frac{1}{2}$. Then there is no symmetric pure strategy equilibrium in which $a_1 = a_2$. There is a symmetric mixed strategy equilibrium in which each buyer gets an expected payoff of 0.*

Proof. (i) Suppose there is a symmetric pure strategy equilibrium with $a_1 = a_2$. Then $a_1 = a_2 \neq 0$, because otherwise some B_i would deviate and set $a_i = a_i^*$. (The surplus from this additional investment would all go to the buyer.) Suppose that $a_i^* > a_1 = a_2 > 0$. This cannot be an equilibrium because some buyer would deviate to a_i^* and do better. Suppose therefore that $a_1 = a_2 = a_i^*$. Since both buyers have the same values, the price to be paid to the seller will be $\beta a_i^* + g(\frac{1}{2})\alpha\gamma^*$ (i.e., the

---

[7]Obviously, changing the assumption might change the result, but the main point of the paper is that if buyers are not too dissimilar, the seller will choose type of investment (inefficiently) to induce competition among the buyers, and the success of this strategy depends on the bargaining procedure. This "punch line" is not affected if we make buyers very dissimilar.

total surplus from the relationship) so each buyer will get a negative payoff because of the cost of investment. Therefore, some player deviating to 0 investment will obtain a higher profit.

(ii) For any mixed strategy equilibrium (symmetric or not), we first note the lower bounds of the supports for both buyers must be the same, i.e., $\underline{a}_1 = \underline{a}_2$. Otherwise, the one whose lower bound is lower will always make a loss by choosing any positive investment strictly below the other buyer's lower bound (as he will never win to recoup its investment cost), and this buyer could have done better by not investing at all. We next claim that the two lower bounds of supports must indeed be zero, i.e., $\underline{a}_1 = \underline{a}_2 = 0$, and as a consequence the expected payoff of each buyer must be zero. Suppose not so that $\underline{a}_1 = \underline{a}_2 > 0$. Then buyer i choosing $a_i = \underline{a}_i$ wins with positive probability only if $a_j = \underline{a}_j = \underline{a}_i$, where $j \neq i$. But the payoff to B_i from such an outcome is still negative $(-c(\underline{a}_i))$ and B_i can gain by setting $\underline{a}_i = 0$. Finally, to calculate the mixed strategy $F_i(\bullet)$, we set up the usual expression for the expected payoff of B_j for any point a_j in B_j's support, where $i = 1, 2$, and $j \neq i$. The expected payoff equals

$$\beta \int_0^{a_j} (a_j - a_i) dF_i(a_i) - c(a_j).$$

Differentiating the expression with respect to a_j and setting the resulting term equal to 0, we obtain

$$\beta F_i(a_j) - c'(a_j) = 0.$$

Therefore

$$F_i(a) = \frac{c'(a)}{\beta}. \tag{6}$$

Since this expression applies to both buyers, $F_1(\bullet) = F_2(\bullet)$ and the mixed strategy equilibrium is indeed symmetric. $\square$

Remark 2. It can be verified from (5) that if $c(a_i) = \frac{1}{2}a_i^2$, $F(a_i)$ is uniform from 0 to β.

We note that, for the seller, γ^* is still the equilibrium strategy, since the payoff to S is $E\{\min\{\beta\tilde{a}_i, \beta\tilde{a}_2\}\} + g(\frac{1}{2})\alpha\gamma - c(\gamma)$, given that $x = \frac{1}{2}$ is still optimal.

Asymmetric equilibria do not exist in this game because of the assumption that buyer investment is less important than the seller's. Consider the following example.

Example 3: We assume that $c(\gamma) = \frac{1}{2}\gamma^2$ as in Example 1. Consider the following candidate (investment) equilibrium strategies:

B_1 : Invest a_1' such that

$$\frac{1}{1+\delta}\beta = a_1'.$$

B_2 : Invest 0.

S : Set $x = 0$ and invest γ' such that

$$\gamma' = \frac{\delta}{1+\delta} \bullet \alpha.$$

Let us take $\delta \to 1$ for this example. Then S's payoff will be

$$\frac{1}{2}[\beta a'_i + \alpha\gamma'] - \frac{1}{2}\gamma^2$$

$$= \frac{1}{2}\left[\beta \bullet \frac{1}{2}\beta + \alpha\frac{1}{2}\alpha\right] - \frac{1}{2}\left(\frac{1}{2}\alpha\right)^2$$

$$= \frac{\beta^2}{4} + \frac{\alpha^2}{4} - \frac{\alpha^2}{8} = \frac{\beta^2}{4} + \frac{\alpha^2}{8}.$$

Now suppose S deviates and chooses x such that $g(1-x) > \frac{1}{2} > g(x)$ and $\beta a'_1 + g(x)\alpha\gamma'' = g(1-x)\alpha\gamma''$, where now γ'' is the level of investment chosen. (Such a x will clearly exist provided a'_1 is not too high).

Given the choice of x, γ'' can be determined by the first-order conditions as before, so that

$$\gamma'' = g(1-x) \bullet \alpha.$$

The seller now gets the entire payoff because B_1 and B_2 bid away their entire surpluses.

Therefore the payoff to S is

$$\beta \bullet \frac{\beta}{2} + g(x) \bullet \alpha(1-x) \bullet \alpha - \frac{1}{2}(g(1-x)\alpha)^2.$$

Subtract from this S's payoff in the candidate equilibrium, namely $\frac{\beta^2}{4} + \frac{\alpha^2}{8}$, to obtain

$$\frac{\beta^2}{4} + \alpha^2\left[g(x) \bullet g(1-x) - \frac{1}{2}(g(1-x))^2 - \frac{1}{8}\right].$$

But we can substitute now for $\frac{\beta^2}{4}$, since

$$\frac{\beta^2}{2} + g(x) \bullet \alpha \quad g(1-x)\alpha = (g(1-x))^2\alpha^2.$$

The difference in payoffs is then

$$\alpha^2\left[\frac{(g(1-x))^2}{2} - \frac{g(x)g(1-x)}{2} + g(x)g(1-x) - \frac{(g(1-x))^2}{2} - \frac{1}{8}\right]$$

$$= \frac{\alpha^2}{2}\left[g(x)g(1-x) - \frac{1}{4}\right].$$

We know that when $x = 0$, $g(x) \bullet g(1-x) = 0$ and likewise for $x = 1$. Symmetry (and the concavity of the product) demonstrate that $g(x)g(1-x)$ reaches its maximum value at $x = \frac{1}{2}$, but since each of the components of the product is greater than $\frac{1}{2}$, the product is greater than $\frac{1}{4}$.

Therefore, there could exist values of parameters such that $g(x)g(1-x) > \frac{1}{4}$, hence a profitable deviation could exist. In this case the candidate equilibrium is not actually an equilibrium.

We have considered only the asymmetric pure strategy profile with S investing at $x = 0$ or $x = 1$. If S optimally invests so as to remove the asymmetry completely, the buyer who invests a positive amount will get a negative payoff and will deviate to 0 investment. The remaining case is the seller partially removing the asymmetry so as to get more than $1/2$ the payoff with a particular buyer without completely removing it. Given the condition in Section 4.2 of the relative unimportance of the buyer's investment compared to the seller's, the seller has an incentive to choose x strictly in the interior to induce a larger outside option (as in the proof of Proposition 1.) Such an asymmetric equilibrium would have the buyer investing optimally but the seller choosing x not equal to 0 or 1, thereby introducing inefficient investment-type choice.

5 Sequential Offers Extensive Form

5.1 The bargaining procedure

We now consider a bargaining procedure where a randomly chosen buyer, B_1 or B_2, makes a proposal to the seller S. The seller can accept or reject. If the seller accepts, the game is over. If S rejects, she makes an offer, but chooses either one of the buyers to bargain with. The chosen buyer B_i then accepts or rejects and so on. Between a rejection and a new proposal, time elapses (as in Rubinstein, 1982) and the common discount factor is δ, as before. The payoffs have already been described in the model section.

This extensive form has been studied, for example in Osborne and Rubinstein (1990) or in Chatterjee, Dutta, Ray and Sengupta (1993) where it is an example in a general study of coalition formation. We summarize the result in a proposition.

Proposition 3. *Let v_1 be the surplus if a $B_1 - S$ trade takes place and v_2 be the surplus if a $B_2 - S$ trade takes place. Let $(v_i - p, p)$ be a (feasible and efficient) agreement between B_i and S where the second entry is the payment from the buyer to the seller while the first entry is the part of surplus left over to the buyer, $i = 1, 2$. The subgame perfect equilibrium is as follows:*

1. *Suppose $v_1 > v_2$ (the case of $v_2 < v_1$ is symmetric and omitted). B_1 always offers $(\frac{1}{1+\delta}v_1, \frac{\delta}{1+\delta}v_1)$ and accepts offers that give him at least $\frac{\delta}{1+\delta}v_1$ and rejects otherwise. B_2 always offers $(v_2 - \frac{\delta}{1+\delta}v_1, \frac{\delta}{1+\delta}v_1)$ if $v_2 > \frac{\delta}{1+\delta}v_1$ and always makes a rejected offer otherwise. B_2 always accepts offers that give him at least $\delta \max(v_2 - \frac{\delta}{1+\delta}v_1, 0)$. S always offers to B_1, with proposal $(\frac{\delta}{1+\delta}v_1, \frac{1}{1+\delta}v_1)$, and always accepts offers that give her at least $\frac{\delta}{1+\delta}v_1$, rejecting otherwise.*

2. *Suppose $v_1 = v_2 = v$, the two buyers always offer $(\frac{1}{1+\delta}v, \frac{\delta}{1+\delta}v)$ and accept any offer that gives him at least $\frac{\delta}{1+\delta}v$. No player (buyers or seller) accepts an offer giving him or her less than $\frac{\delta}{1+\delta}v$.*

Proof. See Osborne and Rubinstein (1990) and Chatterjee *et al.* (1993). □

The striking property of sequential offers is that the seller is not able to reap the benefits of being on the short side of the market. In the "auction-like" mechanism, the bargaining equilibrium is in the core of the game in that the seller obtains the entire surplus. These obviously correspond to different institutions. The "auction-like" mechanism appears to be more appropriate in a "bazaar" setting where buyers make simultaneous offers, while sequential offers seems a better approximation (though by no means a perfect one) to bargains as diverse as those of house sales (in the US) to merger negotiations.

Paradoxically, the sequential offers bargaining might contain the appropriate incentives for investment, more than the auction mechanism described in the previous section. We know turn to this issue.

5.2 Seller investment under sequential offers bargaining

It is clear that there is no incentive for S under this bargaining procedure to choose any value of x other than 0 or 1.

Proposition 4. *If the bargaining procedure used is sequential offers, S will set $x = 0$ or 1 and choose*

$$c'(\gamma') = \frac{\delta}{1 + \delta}\alpha.$$

Proof. The payoff to the seller S is

$$\frac{\delta}{1 + \delta} \max_{x, \gamma}\{v_1, v_2\} - c(\gamma).$$

Since what matters is the maximum of v_1 and v_2, the value of x must be either 0 or 1; any other value of x will multiply α by a factor of $g(x)$ or $g(1 - x) < 1$. Given this, the optimal value of γ is as stated in the proposition, from the first-order conditions, which are also sufficient. □

Therefore, in Examples 1 and 2, where the seller investment decision had $x \in (0, 1)$, the sequential offers bargaining procedure will generate a larger total surplus.

5.3 Buyer investment

We now consider buyer investment in this model. Since one of the buyers is randomly chosen to propose, there is a chance that the buyer for whom the seller has invested is not the first one to propose.[8]

---

[8]Here random choice of the first proposer is more *general* than assuming the buyer for whom the seller has invested makes the first proposal. Modifying the extensive form to have that buyer move first would not affect any substantive results in the paper.

The expected payoff (excluding investment) to B_i, where i is the buyer chosen by the seller is therefore

$$\frac{1}{2}\left(\frac{1}{1+\delta}v_i(\gamma)\right) + \frac{1}{2}\frac{\delta^2}{1+\delta}v_i(\gamma).$$

The other buyer has an expected payoff of 0 and therefore will not invest. For simplicity, suppose $i = 1$.[9]

Proposition 5. *B_1 will invest an amount a'_1 such that*

$$\frac{1+\delta^2}{2}\frac{1}{1+\delta}\beta = c'(a'_1),$$

and B_2 will invest 0.

Proof. Similar to previous propositions. □

Note than in comparison to the "auction-like" mechanism, there is no inefficient over-investment by buyers, such as B_2 investing a positive amount. (If an asymmetric equilibrium exists in the auction mechanism, and under the assumption that S prefers B_1, other things being equal, then this too avoids inefficient overinvestment and elicits more investment from B_1 than under sequential offers (though the difference goes to 0 as $\delta \to 1$).)

6 Discussion about Property Rights and Conclusions

To summarise, this paper has considered three types of inefficiency in the choice of investment in a market setting. The first, well-known as the "hold-up problem", involves a player investing too little because of the surplus division in the *ex post* bargaining. The second, present in our earlier paper and in Cai (2003), is about the type of investment chosen by the seller in order to exploit her market power-generalist or specific. A third, minor, cause of inefficient buyer investment in this paper has to do with the order in which buyers approach the seller. While the first source of inefficiency can be reduced in the usual Hart-Moore way of giving property rights to the seller, mitigating the second cause would require property rights for the seller's asset to be held by one of the buyers (not jointly). These statements hold for the bargaining procedure that gives all the surplus to the short side of the market.

Bargaining and competition play important and interconnected roles in our paper. We do not take the view that all extensive forms can be chosen in all markets; the nature of the industry one is considering imposes constraints on the type of bargaining that can take place. With sequential offers, individual ownership of one's

---

[9]S has lexicographic preference. Other things being equal, she prefers B_1 to B_2. (This is another way to state the assumption that $i = 1$.)

assets does as well as assignment of property rights in enhancing efficiency of investment. Thus, if a natural bargaining procedure is used, arms length relationships can still be sufficient.

As pointed out earlier, the choice of specific versus general human capital investment appears quite frequently in the labour market. An electrical engineer who chooses to pursue an MBA instead of, say, a telecommunications degree, is choosing general investment over specific investment. Some would argue that this is, in fact, inefficient, though there are no evident property rights solutions for that problem. The scope of this paper is therefore broader than the incomplete contracts and property rights framework.

References

Cai, Hongbin (2003), "A Theory of Joint Asset Ownership," *RAND Journal of Economics*, Vol. 34, no. 1, Spring 2003.

Chatterjee, Kalyan and Y. Stephen Chiu, (2000, revised 2006), "When Does Competition Lead to Efficient Investments?" mimeo, School of Economics and Finance, University of Hong Kong.

Chatterjee, Kalyan and Bhaskar Dutta (1998), "Rubinstein Auctions: On Competition for Bargaining Partners," *Games and Economic Behavior*, 23, May 1998, 119–145.

Chatterjee, Kalyan and Bhaskar Dutta, Debraj Ray, and Kunal Sengupta (1993), "A Noncooperative Theory of Coalitional Bargaining," *Review of Economic Studies*, 60(2): 463–77.

Chatterjee, Kalyan and Robert C. Marshall (2003), "A Model of Academic Tenure," mimeo Penn State Department of Economics.

Chiu, Y. Stephen (1998), "Noncooperative Bargaining, Hostages, and Optimal Asset Ownership," *American Economic Review*, 88(4): 882–901.

Cole, Hal, George Mailath and Andrew Postlewaite (2001a), "Efficient Non-Contractible Investments in Large Economies," *Journal of Economic Theory*, 101, 2001, pp. 333–373.

_____ (2001b), "Efficient Non-Contractible Investments in a Finite Economy," *Advances in Theoretical Economics*, Vol. 1 [2001], No. 1, Article 2.

de Meza, David and Ben Lockwood (1998a), "Does Asset Ownership Always Motivate Managers? Outside Options and the Property Rights Theory of the Firm," *Quarterly Journal of Economics*, 113(2): 361–86.

_____ (1998b), "Investment, Asset Ownership and Matching: the Property Rights Theory of the Theory in Market Equilibrium," mimeo, Universities of Exeter and Warwick.

Felli, Leonardo and Kevin W.S. Roberts (2002), "Does Competition Solve the Hold-up Problem?" mimeo, LSE and Nuffield College, Oxford.

Gans, Joshua (2003) "Markets for Ownership," Melbourne Business School Working Paper.

Grossman, Sanford J and Hart, Oliver D (1986), "The Costs and Benefits of Ownership: A Theory of Vertical and Lateral Integration," *Journal of Political Economy*, 94(4): 691–719.

Hart, Oliver and John Moore (1990), "Property Rights and the Nature of the Firm," *Journal of Political Economy*, 98: 1119–1158.

Muthoo, Abhinay (1999), *Bargaining Theory with Applications*, Cambridge University Press, Cambridge and New York.

Osborne, Martin J and Ariel Rubinstein (1990), *Bargaining and Markets*, Academic Press, New York.

Rajan, Raghuram G and Luigi Zingales (1998), "Power in a Theory of the Firm," *Quarterly Journal of Economics*, 113(2): 387–432.

Rubinstein, Ariel, (1982) "Perfect Equilibrium in a Bargaining Mode," *Econometrica*, Vol. 50.

Shaked, Avner (1994), "Opting out: bazaars versus 'hi tech' markets," *Investigaciones Economicas*, 1994, 18(3): 421–432.

Review of Economic Studies (1993) **60**, 463–477   0034-6527/93/00230463$02.00
© 1993 The Review of Economic Studies Limited

A Noncooperative Theory of Coalitional Bargaining

KALYAN CHATTERJEE
Pennsylvania State University

BHASKAR DUTTA
Indian Statistical Institute

DEBRAJ RAY
Boston University and the Indian Statistical Institute

and

KUNAL SENGUPTA
University of California, Riverside

First version received November 1987; final version accepted January 1992 (Eds.)

We explore a sequential offers model of n-person coalitional bargaining with transferable utility and with time discounting. Our focus is on the efficiency properties of stationary equilibria of strictly superadditive games, when the discount factor δ is sufficiently large; we do, however, consider examples of other games where subgame perfectness alone is employed.

It is shown that delay and the formation of inefficient subcoalitions can occur in equilibrium, the latter for some or all orders of proposer. However, efficient stationary equilibrium payoffs converge to a point in the core, as $\delta \rightarrow 1$. Strict convexity is a sufficient condition for there to exist an efficient stationary equilibrium payoff vector for sufficiently high δ. This vector converges as $\delta \rightarrow 1$ to the egalitarian allocation of Dutta and Ray (1989).

1. INTRODUCTION

We explore an extensive-form approach to modelling n-person coalitional bargaining situations. Our formulation assumes transferable utility and perfect information, and represents a natural generalization of Rubinstein (1982). Like his and other alternating-offer models, our analysis implies a commitment to the offer by the proposer until all respondents have accepted or rejected the offer. However, we preclude *all* other forward commitments, whether by individuals or by coalitions.[1]

We limit our analysis to *stationary* perfect equilibria of the coalitional bargaining game. The main justification for this restriction is a result proved in Chatterjee, Dutta, Ray and Sengupta (1990) which showed that *all* efficient and individually rational allocations can be supported as perfect equilibria for high enough discount factors, if history-dependent strategies are permitted. In contrast, the predictions of the model are drastically sharpened by invoking stationarity. It can also be argued that stationary equilibria constitute a computationally simple class, and that, if "learning" or "teaching" is important in a given economic context, the way to capture these phenomena is through enriching the model, rather than through adjusting the solution concept. Moreover, like Selten (1981), Gul (1989) and others, we find stationary equilibria analytically tractable.

1. Gul (1989) analyses an extensive form in which such commitments are implicitly possible. Selten (1981) and Binmore (1985) contain alternative formulations of characteristic function bargaining. See also Harsanyi (1963), Bennett (1988) and Selten and Wooders (1990).

(We realise, of course, that it is difficult to put forward a logically complete case for excluding non-stationary equilibria in general.)

Our main focus is on the *efficiency* of bargaining outcomes, and on the limiting properties of such outcomes as discounting vanishes. In strictly superadditive games, efficiency is equivalent to the twin requirements of (a) no delay, and (b) formation of the grand coalition. However, we present a set of examples (Examples 1–3) showing that stationary equilibria may violate either of the two requirements of efficiency mentioned above. These examples are remarkable given the environment of perfect information.[2] They are a consequence of the intrinsic coalitional structure of our model.

Section 3 of the paper presents a characterization of stationary equilibria without delay (Proposition 2). We also provide a sufficient condition that guarantees the existence of stationary equilibria without delay. (Proposition 3).

Section 4 then examines the conditions under which a stationary equilibrium will be efficient. Our main interest is in efficiency "in the limit" as discounting vanishes.

Our bargaining game is defined relative to a *protocol*, that is, an order of proposers and respondents defined for each coalition. We first analyze the issue of efficiency for *all* protocols. We show (Proposition 4) that such a feature is equivalent to the requirement that the game (N, v) satisfies $v(S)/|S| \leqq v(N)/|N|$ for coalitions S. It is easy to see that such games do not have an intrinsic coalitional nature since the grand coalition has higher average worth than every subcoalition. In other words, games where subcoalitions "matter" must exhibit inefficiency for *some* protocol. In deed, the requirement of efficiency for even *one* order of proposers is quite demanding, an observation implied by Proposition 5. It shows that any limit of efficient stationary equilibrium payoff vectors in a strictly superadditive game must yield a payoff vector which is in the *core* of the characteristic function game. This of course immediately implies that a stationary equilibrium can *never* be efficient for high enough discount factors if the characteristic function game has an empty core. Moreover, even if the game has a non-empty core, we show that a game might exhibit inefficiency for *every* protocol (Example 2). Proposition 6 states a sufficient condition for such a phenomenon to occur in general.

In Section 5, we turn to sufficient conditions ensuring efficiency for *some* protocol. Proposition 7 shows that if a game is *strictly convex*, then for high enough discount factors, every stationary equilibrium involves no delay, and there is some protocol under which the stationary equilibrium is efficient.

Finally, we show (Proposition 8) that the (unique) limit payoff vector Lorenz-dominates every other core allocation.[3] We view such a limit result as an additional contribution to the well-known programme of "achieving" cooperative game-theoretic concepts via the play of a strategic game.

2. THE EXTENSIVE FORM

Let (N, v) be a characteristic function game with $|N| = n$. $\mathscr{S}$ is the set of all coalitions. A *protocol* is an ordering of players, one ordering for each coalition. A *proposal* is a pair (S, y) with $S \in \mathscr{S}$ and $\sum_{i \in S} y_i = v(S)$.

---

2. Of course, in incomplete information models, one can have efficiency failure (see Chatterjee and W. Samuelson (1983), Myerson and Satterthwaite (1988), Admati and Perry (1987)). In complete information models that admit multiple subgame perfect equilibrium paths, delay can be generated by using the multiplicity of paths to "punish" deviators. In our paper, however, it may arise even in an essentially unique equilibrium (given the order of proposers), since the reasons for its occurrence have to do with the coalitional structure and not with the multiplicity of equilibria.

3. In fact, this limit allocation is the *egalitarian allocation* of Dutta and Ray (1989).

The game proceeds as follows. The first player in N under the protocol makes a proposal (S, y). Players in S respond sequentially according to the protocol by saying "Yes" or "No". The first "rejector" becomes the new proposer. If no member of S rejects the proposal, then they leave the game (with allocation y) and the game continues with no lapse in time, with player set $N - S$.

We assume that the formulation of a counterproposal takes one unit of time. Let $\delta < 1$ be the common discount factor applied to this time period.

Each proposal, all acceptances of the proposal and a rejection (if any) count as separate *stages* of the game. A *k-history* h_k, for $k \geqq 2$, is a complete listing of the previous $k - 1$ stages. A (pure) strategy for player i assigns proposals at every h_k at which i is a proposer and responses to ongoing proposals at every h_k at which i is a responder. Noting that player i's payoff is $\delta^{t-1}x$ if i receives x in period t, we have a full game-theoretic formulation. One may define in the standard way the notion of a (subgame) perfect (Nash) equilibrium for this game.

A strategy for player i is *stationary* if the decisions assigned to player i are independent of all features of h_k except possibly the set of players remaining in the game and the ongoing proposal (if any). A *stationary equilibrium* is a perfect equilibrium with each player employing a stationary strategy.

Throughout this paper, we consider only stationary equilibria. The main reason for this is the following proposition, which is a generalization of a result due to Herrero (1985) and Shaked (1986).

Say that a bargaining game (N, v) is *strictly superadditive* if for any S, $T \in \mathcal{S}$ with $S \cap T = \emptyset$, $v(S \cup T) > v(S) + v(T)$.

Proposition 0. *Let (N, v) be a strictly superadditive bargaining game with $|N| \geqq 3$. Then there exists $\delta^* \in (0, 1)$ such that for each allocation x with $x_i \geqq v(i)$ for all i and each $\delta \in (\delta^*, 1)$, there exists a perfect equilibrium with the outcome x.*

Proof. See Chatterjee, Dutta, Ray and Sengupta (1990). ‖

This proposition reflects a problem common to much of dynamic game theory. With history-dependent strategies, one can support a plethora of outcomes. It is difficult, moreover, to make a logically convincing argument to rule out such equilibria.

3. STATIONARY EQUILIBRIUM AND DELAY

Fix any stationary equilibrium. Consider a history such that i has to propose at player set S. By stationarity, i, and everyone else who may move after him on the equilibrium of this subgame, behave the same way, irrespective of the history of arrival at player set S. Consequently, we may write $x_i(S, \delta)$ as the subgame equilibrium payoff to i when i is the proposer with S being the player set remaining. For each i, define $y_i(S, \delta) \equiv \delta x_i(S, \delta)$. Then it should be clear that if some individual j were to receive a proposal (T, z) at player set S, with the property that $z_k \geqq y_k(S, \delta)$ for all $k \in T$ that are yet to respond to this proposal (including j, of course), the equilibrium response of j must be to accept. On the other hand, if the above inequality holds for all k who are to respond after j, but *not* for j, then his response will be to reject the proposal.[4] We will therefore refer to $y(S, \delta) \equiv (y_i(S, \delta))$ as the *equilibrium response vector* at the player set S.

We start with a lemma that holds for any equilibrium response vector.

---

[4]. More than this cannot be asserted at this stage. In particular, it is *not* true that j will accept any proposal that yields him more than $y_j(S, \delta)$. We will return to this issue below.

Lemma 1. *Fix a stationary equilibrium and a history that ends with the player set S. Let $y(S, \delta)$ be the equilibrium response vector. Then, for all $i \in S$, we have*

$$y_i(S, \delta) \geqq \delta \max_{i \in T \subseteq S} [v(T) - \textstyle\sum_{j \in T-i} y_j(S, \delta)]. \tag{1}$$

with the property that (1) holds with equality for all i whose equilibrium strategy as a proposer is to make an acceptable proposal at the player set S.

Proof. Using the discussion above, it is clear that if i has to propose, he can guarantee acceptance by choosing any coalition T such that $i \in T$ and by offering $y_j(S, \delta)$ to every $j \neq i$ in T. This implies that $x_i(S, \delta) \geqq \max_{i \in T \subseteq S} [v(T) - \sum_{j \in T-i} y_j(S, \delta)]$, and using the definition of $y_i(S, \delta)$, we obtain (1).

If (1) holds with *strict* inequality, this means that i has made a proposal (T, z) such that $z_j < y_j(S, \delta)$ for some $j \in S$. Consider the protocol restricted to T and look at the last player in the order for which this inequality holds. This player must reject the proposal, by the discussion above. Consequently, i must make an equilibrium proposal that is unacceptable, and this completes the proof of the lemma. ‖

From Lemma 1, it follows that at the player set N, if for some i, $y_i(N, \delta) > \delta[v(T) - \sum_{j \in T-i} y_j(T, \delta)]$ for all $T \subseteq N$, then player i must make an unacceptable offer causing delay along the equilibrium path. The following example demonstrates that there are bargaining games where it pays a player to make such an unacceptable offer in equilibrium.

Example 1 (*Delay*). This example is due to Bennett and van Damme (1988). $N = \{1, 2, 3, 4\}$, $v(\{1, j\}) = 50$, $j = 2, 3, 4$; $v(\{i, j\}) = 100$, $i, j = 2, 3, 4$ and $v(S) = 0$ for all other $S \in \mathcal{S}$.

We claim that for some player starting the game, this example will yield *equilibrium* delay for all discount factors sufficiently close to unity. We prove this by contradiction. Suppose not. Then for all i, (1) holds with equality. Imposing this, we see that $y_i(N, \delta) = \delta 100/(1+\delta)$ for $i \neq 1$ and $y_1(N, \delta) = \delta[50 - (100\delta/(1+\delta))]$.

So if player 1 starts, he gets $y_1(N, \delta)/\delta$, which dwindles to zero as $\delta \to 1$.

Consider, now, that 1 deviates by picking the coalition $\{1, 2\}$ and offering 0 to player 2. 2 will certainly reject the offer. For in the next period, 2 can obtain $y_2(N, \delta)/\delta$ by proposing, say, the coalition $\{2, 3\}$ and offering 3 $y_3(N, \delta)$. This coalition then leaves the game.

Player 4 thus has no option but to share a payoff of 50 with Player 1, whose payoff now approaches 25 as δ goes to 1. So for large δ, player 1 has a worthwhile deviation, a contradiction.[5]

We will be interested in bargaining games for which stationary equilibria do not exhibit delay. This is our task in the next section.

We will be interested in bargaining games for which stationary equilibria do not exhibit delay. This is our task in the next section.

---

5. One can check that for this particular example, delay occurs in *any* perfect equilibrium (not necessarily stationary) for a range of discount factors, whenever 1 starts the game.

3.1. *No-delay stationary equilibrium*

An interesting class of stationary equilibria that will play an important role in our analysis has the property that after every history, the player who has to make a proposal makes an acceptable proposal. We will call such an equilibrium a *no-delay stationary equilibrium.*[6]

By Lemma 1, characterizing a no-delay stationary equilibrium is then equivalent to obtaining a solution to the following class of simultaneous equations. For any coalition $S \in \mathcal{S}$ and discount factor $\delta \in (0, 1)$, let $m(S, \delta) = (m_i(S, \delta))_{i \in S}$ be the solution to

$$y_i(S, \delta) = \delta \max_{T \subseteq S} v(T) - \sum_{j \in T-i} y_j(S, \delta)]. \tag{2}$$

One can prove the following.

Proposition 1. *For each $S \in \mathcal{S}$ and $\delta \in (0, 1)$, $m(S, \delta)$ exists and is unique.*

Proof. We first establish existence. Let $B \equiv [0, \max_{S \in \mathcal{S}} v(S)]^n$. Define $\phi : B \to B$ by $\phi_i(m) \equiv \delta \max_{T \subseteq S} \{v(T) - \sum_{j \in T-i} m_j\}$, for $m \in B$.

Note that $\phi(m) \in B$. By the maximum theorem, ϕ is continuous. By Brouwer's fixed point theorem, the result follows.

To prove uniqueness, we will use the following lemma.

Lemma 2. *For any player set S, consider a vector $y(S, \delta)$ (not necessarily an equilibrium response vector) such that for all $j \in S$, (1) holds. Moreover, for some $i \in S$, suppose that (1) holds with equality. Then for any T that attains the maximum in (1) and for all $j \in T$, we have $y_j(S, \delta) \geq y_i(S, \delta)$.*

Proof. For player i, we have

$$y_i(S, \delta) = \delta[v(T) - \sum_{j \in T-i} y_j(S, \delta)] \tag{3}$$

and for $j \in T - \{i\}$, we have by supposition,

$$y_j(S, \delta) \geq \delta[v(T) - \sum_{k \in T-j} y_k(S, \delta)]. \tag{4}$$

Adding $-\delta y_j(S, \delta)$ on both sides of (4) and using (3), one has

$$(1-\delta)y_j(S, \delta) \geq \delta[v(T) - \sum_{k \in T-j} y_k(S, \delta)] - \delta y_j(S, \delta) = (1-\delta)y_i(S, \delta)$$

Since $\delta \in (0, 1)$, the result follows. ‖

We now return to the main proof. Suppose, on the contrary, that there are two distinct solutions m and m' to (1). Let $K = \{i \in S \backslash m_i \neq m_i'\}$ and w.l.o.g choose $k \in K$ with m_k such that

$$m_k = \max \{z \backslash z = m_i \text{ or } m_i' \text{ for } i \in K\} \tag{5}$$

Of course, $m_k > m_k'$. Let $T \subseteq S$ satisfy

$$m_k = \delta[v(T) - \sum_{j \in T-k} m_j] \tag{6}$$

Then from the definition of m_k',

$$m_k' \geq \delta[v(T) - \sum_{j \in T-k} m_j'] \tag{7}$$

---

6. One could use a weaker definition: that there must be no delay along the equilibrium path alone, but general results regarding such equilibria appear to be considerably more difficult to obtain.

Now by applying Lemma 2, we have for all $j \in T - k$, $m_j \geqq m_k$. So, if $m_j' > m_j$ for any such j, the definition of k in (5) is contradicted. Thus for all $j \in T - k$, we have $m_j' \leqq m_j$. Combining (6) and (7), we therefore obtain $m_k' \geqq m_k$, contradicting (5). ‖

Let Σ^i be the collection of all stationary strategies for Player i, of the following kind:

(A) If i proposes after a history with player set S, he chooses a coalition T where $m_i(S, \delta)$ is attained and proposes the allocation

$$\left\{ \frac{m_i(S, \delta)}{\delta}, (m_j(S, \delta))_{j \in T-i} \right\}$$

for the coalition.[7]

(B) If i responds to some ongoing proposal (T, z) after a history with set of players $S \in \mathscr{S}$, he accepts the proposal if $z_j \geqq m_j(S, \delta)$ for all $j \in T$ who are yet to respond (including i), and rejects it otherwise.

Let $\Sigma \equiv \prod_{i \in N} \Sigma^i$.

It is easy to see that if any strategy vector $\sigma \in \Sigma$ is an equilibrium, then there is no delay in equilibrium nor in any subgame, and the initial proposer gets $m_i(N, \delta)/\delta$. Indeed, by Lemma 1 and Proposition 1 (the uniqueness of $m(S, \delta)$), the converse is also true. This is easily seen for part (A), using the results just cited.

To see part (B), we will use part (A). Assume that for some proposal (T, z), i is the last respondent. Then clearly (B) is true.[8] Now proceed by induction. Fix an integer $k \geqq 0$. Suppose that (B) is true for all proposals (T, z) and respondents j with m more respondents to follow, where $0 \leqq m \leqq k$. Now consider a proposal (T, z) and a respondent i with $k + 1$ respondents to follow. If $z_j \geqq m_j(S, \delta)$ for all j who are yet to respond (including i), it follows easily from the induction hypothesis and Part (A) that i should accept. If not, then either $z_i < m_i(S, \delta)$ or $z_j < m_j(S, \delta)$ for some j to follow i. In the former case, i is better off by rejecting because he can get $m_i(S, \delta)/\delta$ in the next period. In the latter case, too, he is better off by rejecting! For if he accepts, then a later respondent will reject and offer i $m_i(S, \delta)$. If i seizes the initative by rejecting the proposal, he can get $m_i(S, \delta)/\delta$, which is larger.

These observations are collected together in

Proposition 2.   *Let σ be a stationary equilbrium. Then σ is a no-delay equilibrium if and only if $\sigma \in \Sigma$.*

We have already seen in Example 1 that $\sigma \in \Sigma$ need not be an equilibrium. We now provide a sufficient condition under which the set of stationary equilibria is exactly Σ.

Condition M.   For all S, $T \in \mathscr{S}$ and $\delta \in (0, 1)$, if $T \subset S$, then $m_i(S, \delta) \geqq m_i(T, \delta)$ for $i \in T$.

The reader should observe that (M) is quite independent of superadditivity conditions. It neither implies nor is implied by superadditivity.

---

7. Σ_i contains more than one element if and only if for some player set S there is more than one $T \subseteq S$ that solves the maximization problem in (2).

8. For the purpose of ensuring an equilibrium, equality must imply acceptance.

CHATTERJEE *ET AL.* COALITIONAL BARGAINING 469

Proposition 3. *Suppose that condition M holds for the bargaining game(N, v). Then the set of stationary equilibria is exactly Σ. That is, no stationary equilibrium involves delay in equilibrium in any subgame.*

Proof. We first check that any strategy $\sigma \in \Sigma$ is indeed an equilibrium. Pick $i \in N$ and let all players in $j \in N - i$ use the strategies prescribed by Σ^j. Consider a subgame with player set S with i as the proposer. Note that by playing according to Σ^i, player i can guarantee himself a payoff of $m_i(S, \delta)/\delta$. If he makes an unacceptable offer today, he will receive a present value of $m_i(T, \delta)$, for some $T \subseteq S$. Since $\delta < 1$, by Condition (M), this payoff is strictly less. Thus i will make an acceptable offer, which, given the strategies of the other players, must agree with Σ^i. To check responses, proceed exactly as in the discussion preceding Proposition 2.

Next, we show that every stationary equilibrium belongs to Σ. Consider a stationary equilibrium. Consider a history ending with player set S. Let $y(S, \delta)$ be the equilibrium response vector. It will suffice to show that $y(s, \delta) = m(S, \delta)$. This is clearly true if $|S| = 1$. Suppose, for some integer $1 \leq K < n$, that this is true for all S with $|S| \leq K$. Now if the hypothesis is false for some S with $|S| = K + 1$, then, using Lemma 1, $J \equiv \{j \in S \setminus m_j(S, \delta) < y_j(S, \delta)\}$ must be non empty. Pick $i^* \in J$ such that $m_{i^*}(S, \delta) \geq m_j(S, \delta)$ for all $j \in J$.

By condition (M), we have $m_{i^*}(S, \delta) \geq m_{i^*}(S', \delta)$ for all $S' \subseteq S$ such that $i^* \in S'$. Thus by the induction hypothesis, at the player set S, i^* must be making an acceptable proposal (T, z), where $z_j = y_j(S, \delta)$ for $j \in T - i^*$ and $z_{i^*} = y_{i^*}(S, \delta)/\delta$. Now, $T = (T \cap J') \cup (T - J')$, where $J' \equiv \{j \in S \setminus m_j(S, \delta) > m_{i^*}(S, \delta)\}$. By Lemma 2, we have for all $j \in T - J'$,

$$y_j(S, \delta) \geq y_{i^*}(S, \delta) > m_{i^*}(S, \delta) \geq m_j(S, \delta). \tag{8}$$

Now consider $j \in J'$. Let T' be a set that attains the maximum in (2) (for $i = j$). Then, by Lemma 2 and the fact that $j \in J'$, $m_k(S, \delta) \geq m_j(S, \delta) > m_i(S, \delta)$ for all $k \in T'$. Using (1) for $i = j$, we have, therefore,

$$y_j(S, \delta) \geq \delta[v(T') - \textstyle\sum_{k \in T'-j} y_k(S, \delta)] \geq \delta[v(T') - \textstyle\sum_{k \in T'-j} m_k(S, \delta)]$$

$$= m_j(S, \delta)$$

On the other hand, since $m_j(S, \delta) > m_{i^*}(S, \delta)$, we have from the definition of i^* that $y_j(S, \delta) \leq m_j(S, \delta)$. Combining, we see that $y_j(S, \delta) = m_j(S, \delta)$ for all $j \in J'$. Putting this information together with (8), we have

$$y_{i^*} = \delta[v(T) - \textstyle\sum_{k \in T-i^*} y_k] \leq \delta[v(T) - \textstyle\sum_{k \in T\dagger} m_k] \leq m_{i^*}$$

contradicting our supposition that i^* is in J. ‖

4. STATIONARY EQUILIBRIUM AND EFFICIENCY

An allocation $x \in \mathcal{R}^n$ is *feasible* for (N, v) if there is a partition of N into coalitions $(S_1, \ldots, S_k)$ such that $\sum_{i \in S_j} x_i \leq v(S_j)$ for all $j = 1, \ldots, k$.

An equilibrium is *efficient* if there is no feasible allocation for (N, v) such that every agent receives a higher utility (relative to the equilibrium payoff) in that allocation.

Our interest in this section is to try to characterize the set of bargaining games for which the resulting bargaining equilibrium is efficient (for high enough discount factors). It should be recalled that an important element of our bargaining game was the specification of the protocol. Consequently, one is also interested in knowing how our efficiency results are related to this particular aspect of the game.

4.1. *Efficiency of stationary equilibria for all protocols*

We start with a definition. Say that a game (N, v) is *dominated by its grand coalition* if for all $S \subseteq N$, we have

$$\frac{v(N)}{|N|} \geq \frac{v(S)}{|S|}.$$

Remark. Note that the bargaining game analysed by Rubinstein (1982) and its n-person generalization by Herrero (1985), Shaked (1986) and others fall in this category.

We may now state

Proposition 4. *The following statements are equivalent:*[9]

(a) (N, v) *has the property that for every protocol, there is a sequence of discount factors tending to one and a corresponding sequence of efficient stationary equilibria.*

(b) (N, v) *is dominated by its grand coalition.*

Remark. For these games, the limit equilibrium payoff vector is the allocation obtained by equal division of $v(N)$ (see proof below).

Proof. We first show that (b) implies (a). Assume (b). Consider any stationary equilibrium. Let x_i be the payoff to player i from this equilibrium, for any history where N is the player set and i has to propose. For i to obtain this payoff, there must be a history with player set S and i as the proposer with i making an acceptable proposal (T, z), where $z_i = v(T) - \sum_{j \in T - i} z_j \geq x_i$. By stationarity, the equilibrium response vector at S must be δz_i for i and z_j for $j \neq i$. Applying Lemma 2, we have $z_j \geq \delta z_i$ for all $j \in T$. Thus,

$$x_i \leq z_i \leq \frac{v(T)}{1 + \delta(|T| - 1)} \leq \frac{v(N)}{1 + \delta(n - 1)}$$

with the last inequality holding strictly whenever $T \neq N$ (use (b)).

Thus, i's payoff from any stationary equilibrium is bounded above by $v(N)/[1 + \delta(n - 1)]$, and this is true of *every* $i \in N$. So any proposer i at the player set N can make the acceptable proposal (N, z'), where $z'_j = \delta v(N)/[1 + \delta(n - 1)]$ for $j \neq i$ and obtain $v(N)/[1 + \delta(n - 1)]$. So the infimum of player i's equilibrium payoffs is $\delta v(N)/[1 + \delta(n - 1)]$. This implies that the equilibrium response vector y at the player set N is $y_j = \delta v(N)/[1 + \delta(n - 1)]$. Given (b), it then follows that every player makes an acceptable offer forming the grand coalition at the player set N.

Now we show that (a) implies (b). If (a) is true, then (using the finiteness of protocols) there exists $\delta^* \in (0, 1)$ such that for all $\delta \in (\delta^*, 1)$ and all protocols, each person who starts makes a proposal to the grand coalition which is accepted. For $\delta \geq \delta^*$, using Lemma

---

9. The equivalence holds in the space of games that admit a stationary equilibrium. We conjecture, though, that for every game, a stationary equilibrium exists.

1 and 2, the equilibrium response vector $y(N, \delta)$ is then given by $y_i(N, \delta) = \delta v(N)/[1 + \delta(n-1)]$ and the equilibrium payoff to the proposer is simply $v(N)/[1 + \delta(n-1)]$. Since in equilibrium no proposer wants to pick a subcoalition S, we must have

$$\frac{v(N)}{1 + \delta(n-1)} \geq v(S) - \frac{\delta v(N)}{1 + \delta(n-1)} (|S| - 1).$$

Therefore, for every $S \in \mathcal{S}$,

$$\frac{v(N)}{1 + \delta(n-1)} \geq \frac{v(S)}{1 + \delta(|S| - 1)}$$

Passing to the limit as $\delta \to 1$ in the above expression, we get (b). ‖

Proposition 4 thus shows that the requirement of efficiency for all protocols is extremely demanding. According to our definition, a game is dominated by its grand coalition if the average worth of N is no less than the average worth of any subcoalition. Such games are obviously not ones where subcoalitions matter in any real sense. What happens if we do not insist on such a strong requirement?

4.2. Efficiency for some protocol

The negative result of the previous section motivates a weaker inquiry: is it possible to obtain efficiency for at least *one* protocol? We start our analysis by noting a strong property of efficient stationary equilibria.

Proposition 5. *Let (N, v) be strictly superadditive. Suppose that for some protocol, there is a sequence $\delta^k \to 1$ and a corresponding sequence of efficient stationary equilibrium payoff vectors $z(\delta) = (z_i(\delta))_{i \in N}$, converging to z^* as $\delta \to 1$. Then z^* is in the core of the characteristic function (N, v).*[10]

Proof. Since the equilibrium is efficient, it must be that along the equilibrium path, no player makes an unacceptable offer, i.e., there is no delay along the equilibrium path. Moreover, since (N, v) is strictly superadditive, the first proposer (say player 1) must make a proposal to the grand coalition. Therefore, if the equilibrium response vector is $y(N, \delta)$, the equilibrium allocation is given by $z_j(\delta) \equiv y_j(N, \delta)$ for $j \neq 1$ and $z_1(\delta) \equiv y_1(N, \delta)/\delta$. The result now follows by applying Lemma 1, replacing S by N and taking δ to 1. ‖

Remark. Proposition 4 does not assert the existence of the limit vector of payoffs, though we conjecture that this limit is always well-defined. For the case of no-delay stationary equilibria, the existence of this limit follows trivially from Lemma 3 below.

An immediate corollary of Proposition 4 is that for high enough discount factors, games with empty cores will *never* possess efficient stationary equilibrium for *any* protocol. Indeed, more is true. We now present an example which shows that even for games with nonempty cores, stationary equilibria are inefficient for *every* protocol.

Example 2 (The employer-employee game). $N = \{1, 2, 3\}$, $v(\{i\}) = 0$ for all i, $v(\{1, 2\}) = v(\{1, 3\}) = 1$, $v(N) = 1 + \mu$, $0 \leq \mu < 0.5$, $v(23) = \varepsilon$, positive but "small". This game is strictly superadditive, and it has a nonempty core.

10. This result is not true if the game is just superadditive. See Example 2.

It is possible to check the following:

$$
m_i(S, \delta) = \begin{cases}
\dfrac{\varepsilon\delta}{1+\delta}, & S = \{2, 3\}, \, i \in \{2, 3\}, \\[2ex]
\dfrac{\delta}{1+\delta}, & i \in S, \, S \in (\{1, 2\}, \{1, 3\}), \\[2ex]
\dfrac{\delta(1+\mu)}{1+2\delta}, & S = N, \, i \in N \text{ with } \dfrac{\delta}{1+\delta} < \mu, \\[2ex]
\dfrac{\delta}{1+\delta}, & S = N, \, i \in N \text{ with } \dfrac{\delta}{1+\delta} > \mu. \\[2ex]
0, & S = \{i\}.
\end{cases}
$$

Using this list, one can observe that condition M is satisfied for ε sufficiently small. Consequently all stationary equilibria are no-delay.

Now we discuss efficiency.[11] If δ is "small", i.e. $\mu > \delta/(1+\delta)$, the grand coalition always forms in stationary equilibrium with the proposer receiving $(1+\mu)/(1+2\delta)$. The outcome is efficient. However, with $\mu < \delta/(1+\delta)$, (as it will be for δ close to 1), the opposite happens. In *no* stationary equilibrium does the grand coalition form: either coalition $\{1, 2\}$, or $\{1, 3\}$ forms with the the coalition splitting the surplus 1 equally as $\delta \to 1$. The remaining person receives zero. These equilibria are inefficient, though the game has a nonempty core.[12]

The example can be generalized to yield a weaker sufficient condition for inefficiency. The following lemma, which we also use later, will be needed:

Lemma 3. $m^*(N) \equiv \lim_{\delta \to 1} m(N, \delta)$ *is well-defined, and for each* $i \in N$,

$$
m_i^*(N) = \max_{i \in S \subseteq N} [v(S) - \textstyle\sum_{j \in S-i} m_j^*(N)], \tag{9}
$$

with the property that for every set S^* *that attains the maximum in* (9),

$$
m_j^* \geq m_i^* \quad \text{for all } j \in S^*. \tag{10}
$$

The proof of this lemma uses the following two steps. First, one shows that there is a unique vector $m^*(N)$ satisfying (9) and (10). The proof mimics that of Proposition 1. Second, one can easily check that every limit point of $m(N, \delta)$ (as δ goes to 1) must satisfy (9) and (10). We omit the formal details.

An earlier version of our paper (Chatterjee, Dutta, Ray and Sengupta (1987)) shows how the limit vector $m^*(N)$ may be explicitly computed from the parameters of the

---

11. In a quite different game in which only bilateral trade is allowed, Hendon and Tranaes (1990) find that their equilibrium could involve trading inefficiencies.

12. It might be of independent interest to note that if $\mu = 0$, the set of all perfect equilibria in this game coincides with the set of all stationary equilibria. For this case, Player 1 cannot get more than $1/(1+\delta)$ in *any* subgame perfect equilibrium, however much he (she) wishes to "teach" Players 2 and 3 and they seek to "learn". Of course, in this case there is no inefficiency, but the limit outcome is not in the core of the game. This shows, by the way, that strict superadditivity is essential for Proposition 5.

CHATTERJEE *ET AL.*   COALITIONAL BARGAINING   473

model. We use this limit vector to provide a sufficient condition for inefficiency under every protocol:

Proposition 6. *Let* (N, v) *be a strictly superadditive game such that for every* δ, *the set of stationary equilibria is exactly* Σ. *If* $\sum_i m_i^*(N) > v(N)$, *then there exists* $\hat{\delta} < 1$ *such that for each* $\delta \in (\hat{\delta}, 1)$ *and any protocol, each stationary equilibrium is inefficient.*

The proof follows immediately from Lemma 3 and the definition of Σ.

Proposition 6 raises an interesting question: if $\sum_i m_i^*(N) = v(N)$ for a strictly super-additive game, is it true that for *some* protocol, *some* stationary equilibria is efficient 'in the limit'? This would yield a complete characterization. Unfortunately, the answer is no, as the following example demonstrates:

Example 3. $N = \{1, 2, 3\}$, $v(1) = 1$, $v(2) = v(3) = 0$, $v(\{1, 2\}) = 1 \cdot 8$, $v(\{1, 3\} = 1 \cdot 6$, $v(\{2, 3\}) = 0 \cdot 1$, and $v(N) = 2.4$.

This is a strictly superadditive game and condition M holds. Thus all stationary equilibria are no-delay, and by Proposition 2 belong to Σ. Moreover, it is possible to check that

$$m(N, \delta) = (\delta, 1 \cdot 8\delta - \delta^2, 1 \cdot 6\delta - \delta^2)$$

Thus $\sum_i m_i^*(N) = 2 \cdot 4 = v(N)$. However, it can be checked that as $\delta \to 1$, for no player is $m_i(N, \delta)$ attained at the grand coalition N. So we have inefficiency for large discount factors, under every protocol.

In the next section, we continue our search for an interesting class of games which display efficiency for some protocol.

5. EFFICIENCY OF STATIONARY EQUILIBRIA: STRICTLY CONVEX GAMES

In this section, we intend to show that there exists a class of games for which the following holds: for δ close to 1, (i) all stationary equilibria are no-delay and (ii) there exists a protocol for which some stationary equilibrium is efficient.

A game (n, v) is *strictly convex* if for all S, $T \subset N$, with $S - T$ and $T - S$ non-empty, one has

$$v(S \cup T) > v(S) + v(T) - v(S \cap T). \tag{11}$$

Proposition 7. *Let* (N, v) *be strictly convex. Then there exists* $\delta^* \in (0, 1)$ *such that for* $\delta \geq \delta^*$, *every stationary equilibrium involves no delay. Moreover, there is a protocol and a stationary equilibrium relative to that protocol which is efficient.*

Remarks.

1. Proposition 7 cannot be extended to the class of all convex games (where (11) holds with weak inequality). It can be checked that the game considered in example 3 is convex, but no stationary equilibrium is efficient for high enough discount factors.

2. Even for strictly convex games, the result is not true for all discount factors. In Example 3, the game can be made strictly convex by making $v(N) = 2 \cdot 4 + \varepsilon$, for $\varepsilon > 0$. But as long as ε is not too large, it can be shown that there are intermediate ranges of δ such that efficiency is not obtained for any protocol.

In preparation for the proof of Proposition 7, we define the threshold discount factor δ^*. By (11), there exists $\varepsilon > 0$ such that for all S, $T \in \mathcal{S}$ with $S - T$ and $T - S$ non-empty, we have $v(S \cup T) > v(S) + v(T) - v(S \cap T) + \varepsilon$. Now define $\delta^* \in (0, 1)$ by the condition $(1 - \delta^*)v(N) = \varepsilon/2$. The proof of the proposition will use the following lemmas.

Lemma 4. *Suppose $\delta \in (\delta^*, 1)$. Fix any player set S and let $m_i(S, \delta)$ be attained at S^i, for $i \in S$. Then for all j, $k \in S$ (not necessarily distinct), we have either $S^j \subseteq S^k$ or $S^k \subseteq S^j$.*

Proof. For notational convenience, m_i will denote $m_i(S, \delta)$. Now, if the lemma is false, $S^j - S^k$ and $S^k - S^j$ are both non-empty. Let $S^0 \equiv S^j \cup S^k$ and $S_0 \equiv S^j \cap S^k$. We have

$$m_j \geq \delta[v(S^0) - \textstyle\sum_{i \in S^0 - j} m_i]$$

$$\geq \delta[v(S^j) - \textstyle\sum_{i \in S^j - j} m_i] + \delta[v(S^k) - \textstyle\sum_{i \in S^k} m_i] - \delta[v(S_0) - \textstyle\sum_{i \in S_0} m_i] + \varepsilon$$

$$= m_j + m_k(1 - \delta) - \delta[v(S_0) - \textstyle\sum_{i \in S_0} m_i] + \varepsilon.$$

Now if S_0 is empty, $\delta[v(S_0) - \sum_{i \in S_0} m_i] = 0$ and is non-empty, by the definition of m_i, we have, for some $l \in S_0$,

$$\delta[v(S_0) - \textstyle\sum_{i \in S_0} m_i] \leq m_l(1 - \delta) \leq v(N)(1 - \delta).$$

Combining all this information and using the fact that $\delta \geq \delta^*$, we have $m_j > m_j - v(N)(1 - \delta) + \varepsilon > m_j$, which is a contradiction. ‖

By Lemma 4, for each S, each $i \in S$, and each $\delta \in (\delta^*, 1)$, there is a *unique maximal set $S^i(\delta)$* that attains the value $m_i(S, \delta)$.

Lemma 5. *For $\delta \geq \delta^*$, $m_i(S, \delta) \geq m_j(s, \delta)$ if and only if $S^i(\delta) \subseteq S^j(\delta)$.*

Proof. Drop arguments within parentheses for exposition. The "if" part follows from Lemma 2. To prove "only if", suppose on the contrary that $m_i \geq m_j$ but that $S^j \subset S^i$ (this is the only possibility, by Lemma 4). Then by Lemma 2, $m_j \geq m_i$ and so $m_i = m_j$. Now $m_i = \delta[v(S^i) - \sum_{k \in S^i - i} m_k]$. Using $m_i = m_j$ and the fact that $j \in S^i$, this implies that $m_j = \delta[v(S^i) - \sum_{k \in S^i - j} m_k]$. But this contradicts the fact that S^j is the *maximal* set attaining m_j. ‖

Lemma 6. *For $\delta \geq \delta^*$, the condition (M) holds for the bargaining game (N, v).*

Proof. Fix S, choose any $k \in S$, and let $m_k(S, \delta)$ be attained at its maximal set $S^k(\delta) \equiv S^k$. For any $T \subset S$, let $m_k(T, \delta)$ be attained at its maximal set $T^k(\delta) \equiv T^k$. Write $S^k \cap T^k = A$ and $T^k - S^k = B$.

Using the strict convexity of (n, v), and writing $m_i(S)$ (resp. $m_i(T)$) for $m_i(S, \delta)$ (resp. $m_i(T, \delta)$), we get

$$m_k(S) \geq \delta[v(T^k \cup S^k) - \textstyle\sum_{j \in T^k \cup S^k - k} m_j(S)]$$

$$> \delta[v(T^k) + v(s^k) - v(A) - \textstyle\sum_{j \in T^k - k} m_j(T)]$$

$$+ \delta[\textstyle\sum_{j \in B} m_j(T) + \textstyle\sum_{j \in A - k} m_j(T) - \textstyle\sum_{j \in S^k - k} m_j(S) - \textstyle\sum_{j \in B} m_j(S)]$$

$$= \delta[v(T^k) - \textstyle\sum_{j \in T^k - k} m_j(T)] + \delta \textstyle\sum_{j \in B} (m_j(T) - m_j(S))$$

$$+ \delta[v(S^k) - \textstyle\sum_{j \in S^k - k} m_j(S)] - \delta[v(A) - \textstyle\sum_{j \in A - k} m_j(T)] \qquad (12)$$

Now

$$\delta[v(S^k) - \sum_{j \in S^k - k} m_j(s, \delta)] = m_k(S, \delta) \tag{13}$$

and, because $k \in A$,

$$\delta[v(A) - \sum_{j \in A - k} m_j(T)] \leqq m_k(T) \tag{14}$$

Now for $j \in B$, by Lemma 2, we have $m_j(T) \geqq m_k(T)$. Also, since $B \cap S^k = \emptyset$ and $\delta \geqq \delta^*$, we have by lemma 5, $m_j(S) < m_k(S)$ for all $j \in B$. Therefore

$$\sum_{j \in B} (m_j(T) - m_j(S)) \geqq |B|(m_k(T) - m_k(S)) \tag{15}$$

Using (13), (14) and (15) in (12), we obtain for $\delta \geqq \delta^*$,

$$m_k(S) > m_k(T)$$

which completes the verification of Condition (M). ∥

Proof of Proposition 7. Take $\delta \in (\delta^*, 1)$. By Lemma 6 and Proposition 3, every stationary equilibria is no-delay and belongs to σ. Using Lemma 5, there exists a partition $(S_1, S_2, \ldots, S_K)$ of N such that the following holds: if player j belongs to S_j, then $m_j(N, \delta)$ is attained at $\bigcup_{j' \leq j} S_{j'}$. In particular, if $j \in S_K$, then $m_j(N, \delta)$ is attained at N. So any protocol that has any member of S_K as first proposer will yield an efficient, no-delay stationary equilibrium. ∥

Given this positive efficiency result for strictly convex games, one is naturally led to ask: what does the limit equilibrium payoff vector look like? We already know that a limit payoff vector is well defined and given by $m^*(N)$.

To characterize this limit, it will be necessary to introduce the concept of *Lorenz domination.* Consider two allocations x and y. We will say that x Lorenz dominates y if $\sum_{i=1}^{k} \hat{x}_i \leqq \sum_{i=1}^{k} \hat{y}_i$ for all $k \in N$ with strict inequality holding for some k, where $\hat{x}$ and $\hat{y}$ are permutations of x and y in decreasing order.

Lorenz domination represents a partial ordering which has a well known identification with the notion of "greater equality" in payoff distribution.[13]

We can now state our final result.

Proposition 8. *Let (N, v) be a strictly convex game. Suppose that for some protocol, there is a sequence $\delta^k \to 1$ and a corresponding sequence of no-delay stationary equilibria σ^k, such that σ^k is efficient for all k. Then the equilibrium payoff vector converges to a core allocation which Lorenz dominates every other core allocation.*

Remarks.

1. Of course, Proposition 7 ensures that there exists at least one protocol for which the condition of Proposition 8 is met.
2. Using Dasgupta, Sen and Starrett (1973), one can check that this limit equilibrium allocation also maximizes the symmetric Nash product $\prod_{i=1}^{n} x_i$ subject to the core constraints $v(S) \leqq \sum_{i \in S} x_i$ for all $S \in \mathscr{S}$.

---

13. See for instance Kolm (1969), Sen (1969), and Dasgupta, Sen and Starrett (1973).

The proposition (and indeed, its proof below) presumes that there exists such a core allocation that Lorenz-dominates every other core allocation. For this we appeal to the following

Fact (Dutta and Ray (1989)).   *In convex games, there is a core allocation that Lorenz dominates every other core allocation.*

Proof of Proposition 8.   Given the Fact and Propositions 2 and 5, it is sufficient to prove that the limit equilibrium vector $m^*(N)$ is Lorenz undominated by any core allocation. Suppose, contrary to our claim, there exists x in the core of $v(N)$ such that x Lorenz dominates $m^*(N)$. Without loss of generality, write $m^*(N)$ in decreasing order: $m_1^* \geqq m_2^* \cdots \geqq m_n^*$.

Consider the first index i such that $x_i \neq m_i^*$. Using the definition of Lorenz domination, one can show that $x_i^* < m_i^*$, and, moreover, that $x_j \leqq m_j^*$ for all j such that $m_i^* = m_j^*$ (verification of this is straightforward but tedious).

By Lemma 3, we know that there exists S^i such that $m_i^* = v(S^i) - \sum_{j \in S^i - i} m_j^*$, with $m_j^* \geqq m_i^*$ for all $j \in S^i$. But from the previous observation on x, we therefore have $x_i < v(S^i) - \sum_{j \in S^i - i} x_j$. This contradicts our supposition that x is a core allocation. ‖

6. CONCLUSION

The paper has sought to construct a non-cooperative model of coalitional bargaining under complete information. We have shown that inefficiencies could arise in the form of delay and non-formation of the grand coalition. Moreover, the order in which players move turns out to make a significant difference to the efficiency of the equilibrium.

The limitations of this analysis have to do with the particular extensive form used (a curse common to many bargaining models!) and the use of stationary equilibria. While we personally do not find stationarity unpalatable, we are aware that there is a difference of opinion about this. Further research is needed to determine how this assumption bears up experimentally and also to investigate alternative extensive forms. An extension to NTU games is left for a future paper.

Acknowledgements: We are extremely grateful to Elaine Bennett and Eric van Damme for detailed comments that illustrate an error in a previous version of this paper, and to Ariel Rubinstein for a thorough reading and several suggestions about the exposition. We are also grateful to Ken Binmore, Martin Osborne and Reinhard Selten for helpful comments. Various institutions extended their hospitality at different stages of this work, including the Indian Statistical Institute where we started this research. In particular, Dutta and Ray are grateful for support from the Warshow endowment of Cornell University (summer 1987) and Ray also from CNPq and the Instituto de Matemática Pura e Aplicada, 1989–1990. The Penn State Center for Research in Conflict and Negotiation provided support at various times for Chatterjee, Dutta and Sengupta. Finally, we thank the Editor of the *Review* and two anonymous referees for comments and suggestions that led to basic changes in our exposition and choice of material for publication.

REFERENCES

ADMATI, A. and PERRY, M. (1987), "Strategic delay in bargaining", *Review of Economic Studies*, **54**, 345–364.
BENNETT, E. (1988), "Three approaches to bargaining in NTU games" (mimeo., Department of Economics, University of Kansas).
BENNETT, E. and VAN DAMME, E. (1988), Private communication.
BINMORE, K. (1985), "Bargaining and coalitions", in Roth, A. (ed.), *Game Theoretic Models of Bargaining*, (Cambridge: Cambridge University Press).
CHATTERJEE, K., DUTTA, B., RAY, D. and SENGUPTA, K. (1987), "A non-cooperative theory of coalitional bargaining" (Working Paper 87-13, Department of Management Science, The Pennsylvania State University).

CHATTERJEE, K., DUTTA, B., RAY, D. and SENGUPTA, K. (1990), "A theory of noncooperative coalitional bargaining" (Working Paper 90-24, Center for Research in Conflict and Negotiation, The Pennsylvania State University).

CHATTERJEE, K. and SAMUELSON, W. (1983), "Bargaining under incomplete information", *Operations Research*, **35**, 835-851.

DASGUPTA, P., SEN, A. and STARRETT, D. (1973), "Notes on the measurement of inequality", *Journal of Economic Theory* **6** 180-187.

DUTTA, B. and RAY, D. (1989), "A concept of egalitarianism under participation constraints", *Econometrica*, **57**, 615-636.

GUL, F. (1989), "Bargaining foundations of the Shapley value", *Econometrica* **57**, 81-96.

HARSANYI, J. (1963), "A simplified bargaining model for the n-person cooperative game", *International Economic Review*, **4**, 194-220.

HENDON, E. and TRANAES, T. (1990), "Sequential bargaining in a market with one seller and two different buyers" (mimeo., University of Copenhagen).

HERRERO, M. (1985), "A strategic bargaining approach to market institutions" (unpublished Ph.D. thesis, University of London).

KOLM, S-C.(1969), "The optimal production of social justice," in Margolis, J. and Guitton, H. (eds.), *Public Economics*, (London: Macmillan).

MYERSON, R. and SATTERTHWAITE, M. (1983), "Efficient mechanisms for bilateral trading", *Journal of Economic Theory*, **29**, 265-281.

RUBINSTEIN, A. (1982), "Perfect equilibrium in a bargaining model", *Econometrica* **50**, 97-108.

SELTEN, R. (1981), "A noncooperative model of characteristic function bargaining", in Bohm, V. and Nachtkamp, H. (eds.), *Essays in Game Theory and Mathematical Economics in Honour of Oskar Morgenstern* (Mannheim: Bibliographisches Institut).

SELTEN, R. and WOODERS, M. (1990), "A game equilibrium analysis of thin markets", in Selten, R. (ed.), *Game Equilibrium Models: Vol III. Strategic Bargaining.* (Berlin: Springer Verlag).

SEN, A. (1973) *On Economic Inequality* (Oxford: Clarendon Press).

SHAKED, A. (1986), "Three-person bargaining" (talk at the ORSA/TIMS meeting in los Angeles, April).

How Communication Links Influence Coalition Bargaining: A Laboratory Investigation

Gary E. Bolton • Kalyan Chatterjee • Kathleen L. McGinn

Smeal College of Business, Penn State University, University Park, Pennsylvania 16802, and
Graduate School of Business Administration, Harvard University, Cambridge, Massachusetts 02139
Department of Economics, Penn State University, University Park, Pennsylvania 16802
Graduate School of Business Administration, Harvard University, Cambridge, Massachusetts 02139
gbolton@psu.edu • kchatterjee@psu.edu • kmcginn@hbs.edu

Complexity of communication is one of the important factors that distinguishes multilateral negotiation from its bilateral cousin. We investigate how the communication configuration affects a three-person coalition negotiation. Restricting who can communicate with whom strongly influences outcomes, and not always in ways that current theory anticipates. Competitive frictions, including a tendency to communicate offers privately, appear to shape much of what we observe. Our results suggest that parties with weaker alternatives would benefit from a more constrained structure, especially if they can be the conduit of communication, while those endowed with stronger alternatives would do well to work within a more public communication structure that promotes competitive bidding.
(*Bargaining; Coalition; Communication; Experimental Economics; Game Theory; Negotiation*)

1. Introduction: The Role of Communication in Coalition Bargaining

Coalition bargaining sets in motion a host of phenomena, including mergers, joint business ventures, industrial cartels, trade unions, and consumers' cooperatives.[1] Coalition bargaining is by nature multilateral, and so bargainers need concern themselves not only with what to communicate, but with whom to communicate; the *pattern* of communication matters in a way that is not relevant when the negotiation involves just two parties.

The pattern of communication is influenced, sometimes strongly, by the institutional nature of the negotiation. Multilateral trade treaty negotiations typically link bargainers in ways that allow all private and public discussion (e.g., Sebenius 1984). Real estate buyers and sellers, on the other hand, contract to communicate with each other exclusively through a realtor. The United States and China communicated in the early 1970s through Pakistan, which appeared to parlay its role to its own advantage.[2]

Sometimes, bargainers voluntarily restrict communication. In 1999, the copper industry was significantly altered when Phelps Dodge Corporation managed a hostile takeover of fellow copper producers, Asarco Incorporated and Cyprus-Amax Minerals, Inc. Cyprus and Asarco had earlier announced their

---

[1] Game theory, as originally conceived by Von Neumann and Morgenstern (1953, p. 15), revolved around coalitions: "[T]he decisive exchanges may take place directly between large "coalitions," few in number, and not between individuals, many in number, acting independently. Our subsequent discussion of "games of strategy" will show that the role and size of "coalitions" is decisive throughout the entire subject."

[2] Aumann and Myerson (1988) give the example of Syria and Israel, who had no diplomatic relations with each other, but could each communicate with the United States.

0025-1909/03/4905/0583$05.00
1526-5501 electronic ISSN

MANAGEMENT SCIENCE © 2003 INFORMS
Vol. 49, No. 5, May 2003, pp. 583–598

own friendly two-way merger plan, without having responded to Phelps Dodge's inquiries into a three-way merger (Mining Magazine 1999). In contrast, British Aerospace negotiated the sale of its subsidiary, Rover, to BMW, while carrying on separate and simultaneous conversations with Rover's strategic alliance partner, Honda (Pilkington 1996).

While it seems intuitively clear that the communication pattern has a critical influence on coalition outcomes, there has been little work to document the nature of that influence. An experiment by Murnighan and Roth (1977) compared private and public communication to public, and found an effect for privacy. An independent study by Rapoport and Kahan (1976), however, found no effect.[3] More recently, Valley et al. (1992) studied a buyer-seller transaction in which all communication is passed through a middleman, and found that the type of information available to the middleman influences the settlement.[4]

To investigate the effect of the communication pattern, we conducted an experiment that systematically manipulates the institutional constraints on communication in a three-person coalition bargaining game (as we will see, voluntary constraint also plays a role in what we observe). We compare five communication configurations: one which permits all public and private messages, one in which all messages are necessarily public, and three configurations in which a single bargainer controls the flow of messages. This broad manipulation of the communication structure potentially exposes regularities that might not be apparent in the more narrowly-based manipulations of earlier work.

We benchmark the experiment against one of the few coalition bargaining models that explicitly deals

with communication: Myerson's (1977) cooperative game model.[5] A key innovation of Myerson's (1977) approach is the modification of the coalition characteristic function to take account of the communication configuration. Within this framework, we consider two solution concepts: (1) the Shapley value, the solution concept Myerson (1977) considered, and (2) the core. These are cooperative game models, and so we allow relatively free-form bargaining. Other than a time deadline and the restrictions imposed by the communication configuration, our bargainers are free to send messages pretty much as they please.

We find that varying the communication links causes sharp shifts in both the coalitions that form and the profits earned by individual bargainers. Some findings are surprising. For example, while the two "weak" bargainers benefit from controlling the flow of messages, the "strong" bargainer does not. Weak bargainer control yields the highest frequency of grand coalitions, counter to the conventional intuition that unconstrained communication is most conducive to grand coalitions. We also find, for each configuration save all public, a pattern of equal payoffs across certain types of coalitions and (very) unequal payoff allocation across other types. Neither the Myerson (1977) value nor the modified core fits the data particularly well, although, importantly, the data affirms Myerson's method of handling coalitions lacking a connected communication path.

It turns out, somewhat paradoxically, that once we understand which coalitions equally split the payoff, most of the other regularities follow rather naturally from a trace of the competitive pressures in the environment. The coalitions that tend to settle on an equal split are those with the highest per capita value. The unequal splits then tend to look like immediate competitive responses to the equal split proposals. There is a strong tendency to extend two-person coalition proposals in private, and this appears to act as a strategic friction, inhibiting competitive bidding to a certain extent. Private offers are not possible in the

---

[3] While these and our experiments differ in several respects, our findings on public communication suggest a possible explanation for the disparity in the earlier findings (see §5 and Footnote 15).

[4] Most of the experimental work on coalition bargaining, much of it done during the 1970s, was directed at testing cooperative game theory concepts such as the core and the bargaining set, and did not explicitly deal with communication; Selten (1987) provides an overview. More recently, Diermeier and Morton (2000) investigate the Baron-Ferejohn coalition model, and Bottom et al. (2000) investigate the role of risk in coalition bargaining. Okada and Riedl (2001) investigate the role of reciprocal fairness in coalition formation.

[5] Borm et al. (1992) take a similar approach. Most models of coalition bargaining assume, sometimes implicitly, that bargainers are linked in a way that allows all possible private and public communication.

public treatment, and here the coalition allocations have a more uniformly competitive look.

2. The Experiment: A Game, Communication Configurations, and Two Benchmark Models

There are three bargainers: S, C, and T. The characteristic function is defined by

$$v(SCT) = 100, \quad v(SC) = 90, \quad v(ST) = 70,$$

$$v(CT) = 40, \quad v(S) = v(C) = v(T) = 0.$$

This function implies a unique core outcome (this changes when we add communication considerations), the grand coalition with payoffs $(S, C, T) = (60, 30, 10)$. The Shapley value is $(S, C, T) = (46.67, 31.67, 21.66)$. By these measures, the S bargainer is in the strongest strategic position and T is in the weakest. As we will see, this unequal distribution of strategic strength is a desirable feature given the benchmark models involved.

In our experiments, bargainers conducted negotiations by electronic mail. The five experimental treatments are distinguished by the communication configuration.

(1) *Unconstrained.* A bargainer can send a message or proposal to one or both of the other bargainers.

(2) *Public.* Messages and proposals from one bargainer must go to both of the other bargainers.

(3) *S-controls.* All messages and proposals must originate or pass through S. C and T may send communication to S, but not directly to one another. S is under no obligation to pass a message from one bargainer on to another.

(4) *C-controls.* All communication must pass through C.

(5) *T-controls.* All communication must pass through T.

We discuss the models in the context of the game used for the experiment. The *Myerson value* (Myerson 1977) is the Shapley value for the characteristic function game in which the value of a coalition is set to zero if the set of direct communication links between members fails to form a connected path. Consider, for example, the S-controls treatment: CT can form only

if S agrees to act as a go-between, a service that S has a disincentive to perform. Therefore, we modify the characteristic function v by setting the value of coalition CT to zero. We then take the Shapley value of the modified characteristic function: $(S, C, T) = (60, 25, 15)$.

Table 1 lists the Myerson value allocations for the treatments of the experiment. The prediction for the unconstrained treatment is simply the Shapley value. When moving from the unconstrained to the public configuration, the set of connected communication paths is unaffected, so no change is predicted. Giving a single bargainer control of communication leaves the two noncontrolling bargainers without a connected communication path, eliminating the coalition of noncontrolling bargainers from consideration. This raises the average marginal contribution to the grand coalition of the controlling bargainer, which raises his predicted payoff.

It is difficult to challenge the reasoning behind disregarding coalitions without connected communication paths. We retain this feature, and construct an alternative model by substituting the core in place of the Shapley value. We refer to the resulting model as the *modified core*. The predictions are displayed in Table 2. As with the Myerson (1977) value, the grand coalition is the only coalition predicted, and there is no difference in prediction between unconstrained and public. As with the Myerson (1977) value, giving a single bargainer control of communication effectively eliminates a competitive option of the noncontrolling bargainers. The core predictions are not, however, precise; we can only say that the bargainer who controls communication should receive at least as much as he would in unconstrained.

The Myerson (1977) value and the modified core lead us to expect payoff allocations that are differ-

Table 1 Myerson (1977) Value Predictions

	Payoffs		
Configuration/Treatment	S	C	T
Unconstrained and Public	46.67	31.67	21.66
S-controls	60	25	15
C-controls	35	55	10
T-controls	31.67	16.67	51.66

BOLTON, CHATTERJEE, AND MCGINN
How Communication Links Influence Coalition Bargaining

Table 2   Modified Core Predictions

	Payoffs		
Configuration/Treatment	S	C	T
Unconstrained and Public	60	30	10
S-controls	≥ 60	≤ 30	≤ 10
C-controls	≤ 60	≥ 30	≤ 10
T-controls	≤ 60	≤ 30	≥ 10

ent in many respects. There are also some qualitative similarities. Both concepts exclusively predict grand coalitions, independent of communication configuration. Both imply that removing the ability to privately communicate (as with public) makes no difference. The Myerson (1977) value predicts that gaining control of communication increases the controlling bargainer payoff relative to unconstrained. The modified core implies a somewhat weaker version of this hypothesis, simply excluding the possibility of a loss from controlling communication.

3. Laboratory Protocol

The experiment was run in a computer laboratory at the Harvard Business School. Subjects were recruited through campus flyers and newspapers across five Boston universities.[6] Participation required appearing at a special place and time, and was restricted to one session. The opportunity to earn cash was the only offered incentive. In total, there were 99 subjects, 18 in unconstrained, public, and S-controls, 24 in C-controls, and 21 in T-controls (differences reflect variations in subject showup rates).

Procedures were identical for all treatments. The complete instructions appear in Appendix A. The bargaining game was described in terms of a context intended to make the task transparent (similar to an experiment by Kohlberg and Raiffa reported in Raiffa 1982):

> Each negotiation involves representatives from three cement making companies: the Scandinavian Cement Company (SC), the Cement Corporation (CC), and the Thor Cement Company (*Thor*). The three companies are contemplating a formal merger. Each firm would bring value to the merger greater than its own

[6] Harvard, MIT, Boston College, Boston University, and Tufts.

individual profit because of the synergies that would be realized, but how much extra value depends on the mergers that are formed. The following schedule shows the total profit value of all possible mergers, in a fictional currency called thalers:

Merging Parties	Total Profit of Merger (in Thalers)
SC, CC, and *Thor*	100
SC and CC	90
SC and *Thor*	70
CC and *Thor*	40

The instructions then explained the communication links. S-controls, for example, read

> The rules of communication are as follows: SC may send a message to either *Thor* or CC or to both. CC may send a message to SC but not to *Thor*. *Thor* may send a message to SC but not to CC. *The only way for Thor or CC to get a message to one another is to send it through SC.*

Each bargainer was given the e-mail addresses of only those bargainers with whom the communication configuration allowed him to communicate. Negotiations lasted no longer than 8 minutes, and this was public information. The guiding consideration in establishing this particular limit was a desire to keep the session to 90 minutes. Subjects were free to write what they wished, save information about personal identification. They could, and sometimes did, forward messages. They were not allowed to blind carbon copy messages. Messages were directly passed between subjects. After the experiment, the experimenters inspected all messages. No evidence of breaking the rules was found, nor is there any evidence in subject communication of even considering breaking the rules.

There were five rounds of negotiation per session.[7] To control for endgame effects, subjects were not told the number of games they would play. Subject bargaining roles (SC, CC, or *Thor*) were fixed for all rounds. Reputation effects are a particular concern

[7] There were six rounds in the public treatment. For comparability, we drop the sixth round data from the analysis, although it makes no significant difference to include it. All outcome data can be downloaded at the LEMA website, http://lema.psu.edu/, Penn State University.

in a game where individuals fashion their own messages. The design controls for this by ensuring that no two bargainers negotiate together more than once. The necessary rotation scheme is more complex than it is for a two-person game (Appendix B).

We used the simplest formal device that we could think of to certify an agreement: Each party to the agreed-upon coalition completed, either during or immediately after the negotiation, a contract indicating the coalition and the distribution of earnings. An agreement was valid only if the records of the member parties matched and total profits did not exceed the coalition's value. The monitor verified the contracts at the end of the session. Writing down anything but a mutually agreed-upon contract has a negligible chance of matching up.

Bargainers were given a history form to keep a record of their negotiations, and each bargainer had access to the messages she had sent and received in previous games. A bargainer received no information about the outcome of a negotiation beyond what he personally observed.

Each treatment was run as a single session, with the same monitor attending all sessions. Subjects who indicated they knew one another were assigned identical roles. Partner identities were anonymous and remained so after the session. There was a brief tutorial on the use of the e-mail system. The finish time of the negotiation was written on a black board, and a wall clock was visible to all. The monitor verbally announced when 1 minute was left. To avoid wealth effects, subjects were paid for just one negotiation, selected by random lottery after all negotiations were complete. Payments were privately made. Thalers were redeemed in cash at a rate of $0.50 each.[8]

4. Data Analysis

Because our interest is in the comparative influence of the communication configuration, we focus on patterns across treatments. These fall into four categories: (1) coalition formation, (2) communication

control, (3) coalition allocations, and (4) round effects. As will be apparent, the data look substantially different from the benchmark models; we discuss explanations in the next section. For inference tests, we begin with aggregate data, ignoring any factors save for treatment. We then redo the tests restricting attention to last or first round, a procedure that controls somewhat better for heterogeneity, but sacrifices some statistical power. Finally, we present analyses controlling for experience effects with regressions that block for round.[9]

4.1. Coalition Formation

Key Observations. *Coalitions that lack a connected communication path never form. SC is modal in all treatments save T-controls. Grand coalitions are most frequent in T- and C-controls.*

Table 3 displays the frequency of each coalition by treatment. Observe that coalitions lacking a connected communication path never occur. *SC* is modal in all cases save *T*-controls. The incidence of *CT* is always negligible. There are impasses in 4 of the 5 treatments.

A common intuition is that the grand coalition is most likely when communication is least constrained. From Table 3, however, the incidence of grand coalition is lowest in unconstrained (10%), and highest in *T*-controls (80%). Nor does a strong coordinating force necessarily encourage grand coalition formation: *S*-controls had the second lowest incidence of grand coalition (17%). Contingency table tests on the proportion of grand coalitions across treatments (data

---

[8] Average earnings per subject was $21.40 (includes a $10 showup fee), about $14.25 per hour, with a standard deviation of $10.49. The highest paid was $45, while 39 subjects earned only the $10 showup fee.

[9] Even so, the analysis does not control perfectly for the historical contagion inherent to the round-robin design; that is, for any round beyond the first, some bargainers will have had common partners in their history and, hence, the bargaining units are not technically independent. A design that yields unquestionably independent bargaining units would require nearly 500 subjects just to get 33 observations per treatment, and even this is insufficient if we want bargainers to experience multiple games. An alternative approach, common to the two-person bargaining literature, is a series of smaller scale experiments, conducted by independent investigators. Confidence that contagion is not a problem grows with replication, and for the same money that would go into one large-scale experiment, we get a robustness check against any particular set of lab procedures.

Table 3   Coalition Formation

Treatment	$n^\dagger$	Coalition frequency (as % of row)				
		SCT	SC	ST	CT	Impasse
Unconstrained	30	0.10	0.50	0.23	0.03	0.13
Public	30	0.30	0.53	0.03	0.03	0.10
S-controls	30	0.17	0.33	0.27	—	0.23
C-controls	40	0.35	0.45	—	0.03	0.18
T-controls	35	0.80	—	0.20	—	—
Average	33	0.36	0.36	0.14	0.02	0.13

$^\dagger n = 5$ rounds $\times (N/3)$ games per round. There were 18 subjects in unconstrained, public, and S-controls, 24 in C-controls, and 21 in T-controls.

aggregated across rounds) support these observations.[10] Pairwise comparisons of unconstrained with either C- or T-controls reject equality (in both cases, $(p < 0.010)$, and with public weakly reject $(p = 0.058)$; there is no significant difference between unconstrained and S-controls $(p = 0.335)$. For first round data, comparison of unconstrained with T-controls is significant $(p < 0.029)$, and for last round data, comparison of unconstrained with both C- and T-controls is significant $(p < 0.001$ for T-controls and $p = 0.058$ for C-controls). Other comparisons with unconstrained are not significant $(p > 0.200$ in all cases).

4.2. Controlling Communication

Key Observations. *Relative to unconstrained, C- and T-bargainers gain on average from controlling communication, but there is no clear advantage to S-bargainers.*

Table 4 displays, by treatment, average per game payoffs by bargainer type and average total per game payoff. The table also indicates the results of mean test comparisons with unconstrained. Again, the intuition that unconstrained communication is most likely to promote efficient outcomes is not borne out; in fact, T-controls has the highest average per game total payoff $(p = 0.005)$. Rank ordered, unconstrained is second to last in efficiency; only S-controls is less

[10] All contingency table p-values we report are calculated from a simulation of the actual distribution associated with the table (5 simulations of 5,000 iterations each). The procedure avoids potential difficulties associated with the chi-squared approximation of the distribution, for instance when there is a cell with low expected value.

Table 4   Average per Game Payoffs

Treatment	n	Average payoff (standard error)			
		S	C	T	Total
Unconstrained	30	39.9	26.5	6.2	72.7
		(3.42)	(3.84)	(1.68)	(5.77)
Public	30	43.3	30.8	7.5	81.7
		(3.59)	(2.58)	(2.19)	(5.47)
S-controls	30	37.5	21.5	6.3	65.3
		(3.95)	(4.01)	(1.61)	(6.96)
C-controls	40	35.3	38.2**	3.0*	76.5
		(2.87)	(2.89)	(0.68)	(5.85)
T-controls	35	33.0**	24.2	36.8**	94.0**
		(0.52)	(2.13)	(0.54)	(2.06)

*Indicates difference with corresponding unconstrained column value at the 0.10 level (2 tail).
**Indicates difference with corresponding unconstrained column value at the 0.05 level (2 tail).

efficient. These findings hold for each round individually considered, although the differences in average total payoffs are not significant in any single round.

We regressed total payoff on round and dummies for the treatments (omitting unconstrained). The overall regression was significant $(F_{(5, 138)} = 3.538, p = 0.005)$ but accounted for only 10% of the variance in total payoff. The coefficient for round was not significant $(b = 0.108$; t-test $= -1.435$, $p = 0.153)$.[11] T-controls had significantly higher total payoffs than did unconstrained.

Turning to individual payoffs, we see from Table 4 that C and T benefit from controlling communication, but S does not. Restricting attention to the first round, only T significantly benefits from controlling communication $(p < 0.001)$, but by the last round, both C and T benefit $(p = 0.033$ and $p < 0.001)$.[12]

[11] In separate regressions, by treatment, of round effects on total profit, round significantly affects total profit only when T-controls the communication $(F(1, 33) = 4.714$, $p = 0.037$; $R^2 = 0.125)$.

[12] We also ran separate regressions for S's profit, C's profit, and T's profit. The factors in each case were round and dummies for the treatments, omitting the treatment in which the key bargainer controlled the communication. For example, in the test of effects on S's profit, the omitted treatment was S-controls. This allows a simultaneous comparison of the effect on S's profit when S-controls communication, relative to all other communication treatments.

We also looked at average payoffs excluding games that ended with impasses. The results are similar: we find that public, C-controls, and T-controls all have higher average total payoffs than unconstrained ($p < 0.050$); and C and T gain from controlling communication ($p < 0.050$), but S does not (average payoff of 48.9 in S-controls versus 46.1 in unconstrained).

4.3. Coalition Allocations

Key Observations. *In all treatments save public, the path-connected coalition that has the highest per capita value exhibits near equal-division payoffs, and all other two-person coalitions exhibit unequal-division payoffs. In public, there is no clear pattern.*

Table 5 displays average payoffs by type of coalition. The most striking feature is the mix of nearly equal-division and clearly unequal-division allocations. The one exception is the public treatment, where most two-person coalitions tend to look competitive. (Later, we will argue that the public pattern provides important insight into the process behind coalition formation.) Of course, averages can be misleading. Examining individual outcomes leads to the same conclusion: 73% (11/15) of SC coalitions split exactly 45-45 in unconstrained, 90% (9/10) in S-controls, and 67% (12/18) in C-controls versus 25% (4/16) in public.

Figures 1 and 2 offer yet a third way of looking at the allocations. The medians of the distributions are quite similar to the averages in Table 5. The distributions are for the most part quite tight; more often than not, the entire range is 10 units or less. The one exception is public where the variance of S- and T-payoffs in the grand coalition, and

The overall regression for T-profit was significant ($F(5, 159) = 93.129$, $p < 0.0005$). The coefficient for round was positive and significant ($b = 0.113$, t-test $= 2.826$, $p = 0.055$). Each of the coefficients for the four treatments was negative and significant (all $b < -0.750$; all t-tests < -15.000, all $p < 0.0005$). Similarly, the overall regression for C-profit was significant ($F(5, 159) = 3.845$, $p = 0.003$). The coefficient for round was not significant ($b = -0.131$, $p = 0.075$). Each of the coefficients for all treatments was negative and significant (respectively, $b = -0.373$ for unconstrained; $b = -0.291$ for public; $b = -0.346$ for T-controls; and $b = -0.408$ for S-controls, all $p < 0.001$). The overall regression for S-profit was not significant ($F(5, 159) = 1.834$, $p > 0.1$).

Table 5   Average Payoffs Within Coalitions

Treatment	Coalition frequency[†]	Payoffs (standard error)[††]		
		S	C	T
Unconstrained	0.58 (15)	46.7 (1.16)	43.3 (1.16)	
	0.27 (7)	51.9 (1.65)		18.1 (1.65)
	0.04 (1)		30	10
		—		—
	0.12 (3)	45 (2.89)	38.3 (4.41)	16.7 (6.67)
Public	0.59 (16)	54.1 (1.89)	35.9 (1.89)	
	0.04 (1)	35		35
		—		—
	0.04 (1)		20	20
		—		—
	0.33 (9)	44.5 (3.53)	36.6 (2.00)	18.9 (3.79)
S-controls	0.43 (10)	45.5 (0.50)	44.5 (0.50)	
	0.35 (8)	52.5 (3.00)		17.5 (3.00)
	0.22 (5)	50 (0)	40 (0)	10 (0)
C-controls	0.55 (18)	43.5 (0.64)	46.5 (0.64)	
	0.03 (1)		30	10
		—		—
	0.42 (14)	44.9 (1.21)	47.2 (1.06)	7.9 (0.79)
T-controls	0.20 (7)	33.6 (0.92)		36.4 (0.92)
	0.80 (28)	32.9 (0.61)	30.3 (0.61)	36.9 (0.64)

[†]Measured as a proportion of coalitions formed (actual number in parentheses).
[††]A blank indicates that the bargainer is not a member of the coalition.

S- and C-payoffs in two-person coalitions, is higher than in other treatments. Note also the clear difference in payoffs for all coalitions in T-controls versus analogous coalitions in the other treatments.

4.4. Round Effects

Key Observations. *Impasses decrease somewhat, and grand coalitions increase somewhat with experience.*

BOLTON, CHATTERJEE, AND MCGINN
How Communication Links Influence Coalition Bargaining

Figure 1　Boxplot of Bargainer Payoffs by Treatment: Grand Coalitions Only

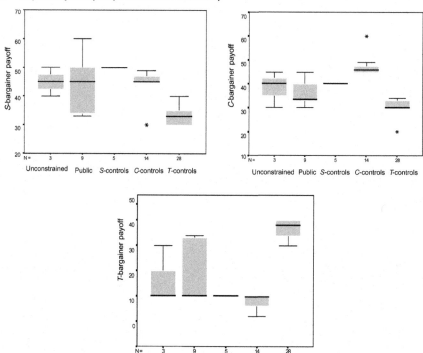

50% of cases have values within the box.
Solid lines within box: median.
Top of box: 75th percentile, bottom of box: 25th percentile.
Top and bottom lines: largest and smallest observed values that are not outliers.
*Extremes: Values > 3 box-lengths from 25th or 75th percentile.

Figure 3 provides a look at the aggregate trend in coalition formation (trends for individual treatments are all quite similar). To see whether grand coalition formation varies round by round, we ran a logistic regression in which the dependent variable is whether the grand coalition formed and the independent variables are a round number and a dummy for each communication treatment save unconstrained. The regression finds a mild but significant round effect ($b = 0.276$, s.e. $= 0.139$; Wald $= 3.941$, $p = 0.047$),

indicating that grand coalitions are somewhat more likely in later rounds. As with the contingency table tests, the regression shows that grand coalitions are more likely in C-controls and T-controls than in unconstrained. The differences are not significant for public and S-controls.

We also find that impasse is less likely in later rounds. In a logistic regression where impasse is regressed on round and dummies for four communication treatments (omitting unconstrained), only the

Figure 2   Boxplot of Bargainer Payoffs, by Treatment: (*SC*) and (*ST*) Coalitions Only

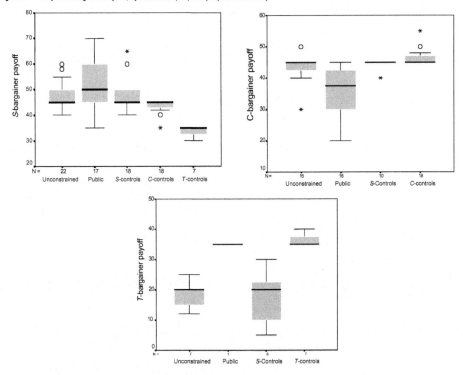

50% of cases have values within the box.
Solid lines within box: median.
Top of box: 75th percentile, bottom of box: 25th percentile.
Top and bottom lines: largest and smallest observed values that are not outliers.
O Outliers: Values > 1.5 box-lengths from 25th or 75th percentile.
Extremes: Values > 3 box-lengths from 25th or 75th percentile.

round coefficient is significant ($b = -0.362$, s.e. $=$ 0.182; Wald $= 3.980$, $p = 0.046$). None of the other coalition formation trends are clearly significant.

5.   Discussion: Explaining the Data
Here, we first consider the two models outlined in §2. Neither is satisfactory, although the modified core fits somewhat better than the Shapley value. So we

focus on why the modified core falls short. It turns out that one observation—the pattern of equal and nonequal coalition payoff allocations—is critical in the sense that if this observation were explained, we would almost surely be able to explain most of the other deviations in the data. One important contributing factor behind the critical observation appears to be the pattern of *voluntary* communication among the bargainers.

Figure 3   Bargaining Outcomes (All Treatments)

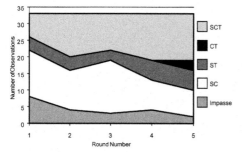

5.1. Comparing the Data with the Models

Table 6 compares observed coalitions with those predicted by each model, allowing for a 5-unit error (an exact match with a point prediction would be extremely demanding). For unconstrained and public, we count any coalition that gives each member its core share as a success for the core; in these treatments, the core is unique, giving players less incentive to agree upon it; e.g., S and C can get $(60, 30)$ on their own. Nevertheless, core logic implies that the distributions of these coalitions should reflect core shares.

The one clear success for these models is that, consistent with Myerson's (1977) method of accounting for the communication configuration, coalitions lacking a connected communication path never occur. Beyond this, the Shapley value has little predictive power. The core's hit rate is highly variable. In C- and T-controls, the core is consistent with a wide variety of grand coalitions (Table 2). But, in fact, grand

Table 6   Coalitions Matching Theory As a Percentage of Coalitions That Form (Within 5 Units[†])

Configuration/treatment	Shapley value	Core
Unconstrained	0%	19% (5/26)*
Public	4% (1/27)	37% (10/27)*
S-controls	0%	0%
C-controls	3% (1/33)	42% (14/33)
T-controls	0%	80% (28/35)

[†]A coalition is "within 5 units" if it is possible to realize the prediction by rearranging no more than five thalers of the actual division.

*Counts any coalition that gives each member his core share as consistent with theory.

coalitions are highly clustered about 45-45-10 and 35-30-35, respectively (Table 5 and Figure 1). The core gives us no reason to distinguish these particular outcomes. In light of this, the core is arguably most successful in the public treatment. Note, however, that the core does not capture any of the key observations discussed in §4.

5.2. Why the Modified Core Falls Short

The logic behind the core suggests that noncore proposals should be blocked by competitive counterproposals. It turns out that most of the nonequal division coalitions we observe, outside of the public treatment, *are* direct competitive counterproposals to the equal-division coalitions. But, apparently, the process of competitive counterproposal goes no further. A striking fact, culled from the bargaining transcripts, is that in approximately half of the games in all treatments save T-controls, the first two-way split proposed is accepted.[13]

Recall that in §4, we found that the coalition with highest per capita value and payoffs split equally is a common settlement in all but the public treatment. The initial attraction to this coalition is quite intuitive. As one SC explained to CC:

> A contract between SC and CC can yield us the greatest profits each. If we split it, we get 45. Any other contract would not yield that much unless one of the parties took a smaller profit than the others.

This is the modal rationale discussed by the bargainers in unconstrained and S-controls, and is mentioned in the bargaining transcripts of all treatments, save for public where it is never mentioned.

It is straightforward to derive the direct competitive responses to the highest per capita/equal-division coalitions. Consider, for example, the unconstrained treatment. (S, C) split $(45, 45)$ is the highest per capita/equal-division coalition. As to competitive counterproposals, T might lure away S by offering a bit more than 45; that is, offering (S, T) split $(45 + \varepsilon, 25 - \varepsilon)$, where ε is the incentive offered to S. Inspection shows that the only other possible direct competitive

---

[13] 44% public; 65% unconstrained; 42% C-controls; 48% S-controls; and 11% T-controls. In T-controls, the first three-way split suggested is accepted in 60% of the games.

response is $(45 + \varepsilon, 45 + \varepsilon, 10 - 2\varepsilon)$.[14] How this particular coalition comes about is nicely illustrated in a message one T sent to S and C

> Listen, I understand if SC and CC merge, you each can get 45. All I ask is to keep me involved. I'll take 10 and you can both keep 45.

Deriving the direct competitive responses for all treatments (after setting coalitions lacking a connected communication path to zero), we arrive at the sets of coalitions listed in the middle column of Table 7. The last column compares with the data, using the same 5-unit criterion used in Table 6 (this allows for both a small ε as well as a small error term). The fit is pretty good in every treatment save public. (And the only type of coalition that is posited but not observed is (C, T) split $(5 - \varepsilon, 35 + \varepsilon)$ in T-controls.)

Most of the key observations from the data are also anticipated in Table 7. The same set of coalitions is implied for S-controls as for unconstrained and public, consistent with the finding that S gains little from controlling communication. C always gets at least 45 from control, and is never left out of a coalition, and T gets at least 35 from control, more than is possible in any other treatment; it is not much of a jump to the conclusion that C and T should gain from control. Also in C- and T-controls, bargainers have a clear incentive to pursue the grand coalition because the amount of additional surplus to divide (10 and 30, respectively) is greater in the grand coalition than in any alternative. In the other treatments, supposing S receives 45 in coalition with C, the amount of additional surplus S and T have to split in their two-person coalition (25) is greater than in the grand coalition (10), so bargainers have less incentive to form grand coalitions than in C- and T-controls.

The one treatment where the Table 7 fit is poor is public. Here, outcomes should look like those for unconstrained and S-controls but, in fact, are different on just about every dimension. Notably, there are few equal-division coalitions among public settlements.

Observe that if we consider more than one competitive reaction, none of the coalitions listed in Table 7

[14] Rapoport and Kahan (1976) observed a similar pattern of grand coalition formation in their experiment.

Table 7   Highest per Capita/Equal-Division Coalitions and Competitive Counteroffers

Configure/treatment	Payoffs[†]			Agreement with data*
	S	C	T	
Unconstrained, Public and S-controls	45	45		U: 85% (23/26),
	$45 + \varepsilon$		$25 - \varepsilon$	P: 59% (16/27),
	$45 + \varepsilon$	$45 + \varepsilon$	$10 - 2\varepsilon$	S: 96% (22/23)
C-controls	45	45		85% (28/33)
	$45 + \varepsilon$	$45 + \varepsilon$	$10 - 2\varepsilon$	
T-controls	35		35	94% (33/35)
		$5 - \varepsilon$	$35 + \varepsilon$	
	$35 + \varepsilon$	$30 - 2\varepsilon$	$35 + \varepsilon$	

[†] A blank indicates the bargainer is not a member of the coalition.

*Counts any coalition that gives each of its members his core share as consistent with theory (allowing for 5-unit error).

would continue to be stable. In the first three treatments listed, for example, $(45 + \varepsilon, 45 + \varepsilon, 10 - 2\varepsilon)$ is vulnerable to (S, T) splitting $(56, 14)$ for all feasible values of ε.[15] The main point to be taken away from Table 7 is that if a theory can explain why competitive counteroffers sometimes fail to be made against the highest per capita/equal-division coalitions and almost always fail to be made against direct competitive responses to the highest per capita/equal-division coalitions, then the theory will also probably explain many of the regularities in our data.

5.3. Private Communication and the Public Treatment

While our data does not permit a definitive explanation for why competitive counterproposals are so limited, we are able to identify some probable contributing factors.

Social equality norms are one probable contributing factor. These norms are known to blunt the kind of competitive responses that the core implies.[16] That said, norms cannot be, in and of themselves, a complete explanation. For one, they do not account for

[15] If we go further and allow all possible competitive reactions, we arrive back at the modified core.

[16] There is substantial evidence for the influence of these sorts of norms in bargaining and other games, see for example, Binmore et al. (1993), Roth (1995), Van Huyck et al. (1995), and Van Huyck et al. (1997).

BOLTON, CHATTERJEE, AND MCGINN
How Communication Links Influence Coalition Bargaining

the many nonequal division settlements we observe. Second, the transcripts suggest this was not the driving factor for many of the equal splits: Fairness was mentioned as a rationale for offers in one-third of the games in the public treatment, where the payoffs are the most unequal, but was never mentioned in the unconstrained transcripts, and is mentioned no more than two times in any of the treatments in which one party controls communication.

Another probable factor has to do with how bargainers communicate their offers. The relevant clue in the outcome data is the distinctive outcomes in public. A natural conjecture is that these settlements look more competitive because public communication induces more iterations of competitive counterproposals.

When we analyze the bargaining transcripts, a striking pattern jumps out: In *every* case in every treatment in which it was permissible (that is, every treatment save for public), a two-person coalition proposal was *privately* communicated to the proposed-to bargainer. In all of the treatments other than public, there were a total of four cases in which a counterproposal was offered in response to an initial two-way split proposal. In contrast, in nearly one-third of the public cases (8/27), proposals and counterproposals were offered throughout the entire exchange, turning it into a veritable haggling session.

There are several reasons to think that a propensity for private communication would inhibit competitive counterproposals. For one, private offers deprive the excluded bargainer of information relevant to proposing a competitive counterproposal. Second, private conversation may be more conducive to eliciting commitments. There is evidence that people tend to abide by their commitments even when there is nothing to bind them (Cialdini 1993). So members of these coalitions may be less likely to go looking for competitive counterproposals.[17]

Figure 4   Time to Agreement

	Average (minutes)
Unconstrained	6.8
Public	7.6
S-controls	6.4
C-controls	6.8
T-controls	6.7

Evidence that private offers lead to a different negotiating dynamic is also found in the time-to-agreement data displayed in Figure 4. A contingency table test finds that only public, where no private offers are possible, is significantly different from other treatments ($p = 0.023$). Fully, 70% of public settlements are reached in the last minute of negotiation (reflecting the "haggling" nature of these interactions), compared to an overall average of 39% for the other treatments. So when all conversations are public, bargainers tend to negotiate straight to the deadline, consistent with the notion that public communication creates a more competitive atmosphere.[18]

---

[17] Murnighan and Roth (1977) found that allowing private messages tended to move coalition divisions closer to equal division, whereas Rapoport and Kahan (1976) found that it had no effect. The rules of the bargaining game studied by Rapoport and Kahan (1976) required that an agreement be made public prior to ratification, independent of whether the agreement had been arrived at by private messages. Any bargainer left out of the tentative agreement was then given a chance to make a counterproposal. Hence, Rapoport and Kahan's (1976) game guaranteed an opportunity for competitive overbidding and undercutting even in treatments with private messages. No such mechanism was available in the bargaining game studied by Murnighan and Roth (1977).

[18] One might think that bargainers find the 8-minute time limit more logistically restrictive in the public treatment. But then we would expect impasses to be positively correlated with time to settlement. In fact, from Table 3, public treatment impasses are below average, even though the deadline effect is largest in this treatment.

While we cannot rule out that more experience would lead to increased competitive bidding in the nonpublic treatments, there are reasons to doubt it. For one, the experience that subjects were given, five games, showed but a small experience effect. Moreover, deviating from the coalitions prescribed in Table 7 is risky. In the public treatment, the S-bargainer gets no more than 45 (and sometimes much less) in 59% of the games (Table 4). So if bargainers are even a bit risk averse, the equal-division deal may be optimal.[19] Evidence that subjects are mindful of the risk is found in the transcripts. In approximately one-third of all the games across treatments,[20] the risk of ending with no deal or being left out of a deal, is mentioned explicitly. Consider, for example, this communication from SC to CC:

> I think it would be a safer situation for us to sign this agreement than to pursue individually with *Thor*. If we each go for a better one with *Thor*, we each have a 50-50 shot of getting screwed out of any money, whereas this deal is a safe 45 each.

Even the weak player recognizes the risk inherent in pushing too hard or for too much, as seen in this response from *Thor*:

> I will settle for 10 since otherwise I will not exist.

5.4. Toward a More Formal Model

It is hard to see how our bargainers can arrive at core solutions absent the free exchange of proposals and counterproposals. Private communication is one inhibiting factor, and there are conceivably others that we have not been able to detect. The general point is that the *process* of negotiation is critical to what we observe. It seems to us that further inquiry into the matter requires a more carefully worked out theory to

guide the investigation. Here, we point in a direction that such a theory might come from.

In principle, some, if not most, of what we have observed might be captured by extending existing noncooperative models of coalition bargaining. Non-cooperative models explicitly deal with the process of negotiation. For example, Chatterjee et al. (1993) model an n-person negotiation as a sequence of proposals and counterproposals. A proposal consists of a coalition and a division of the coalition payoff. Once the proposal is on the table, members of the named coalition sequentially respond, indicating either acceptance or rejection. If everyone accepts, the proposal is adopted and the game ends, otherwise, the first rejecter makes a counterproposal. Individual payoffs are discounted between a rejection and a counterproposal at a common rate of δ, and utility is transferable. The game has an infinite horizon. The model is solved by stationary subgame perfect equilibrium (stationary in the sense that bargainer proposals and responses are independent of play in past rounds) in the limit as $\delta \to 1$, a situation where the amount of time discounting between proposals is "slight."

It turns out that the highest per capita value coalition dividing equally is a general characteristic of the solution to this model (see Chatterjee et al. 1993). In fact, if the value of all coalitions lacking a connected communication path is set to zero, the predicted coalitions and associated allocations are similar to those in Table 6.[21] Why doesn't competition undermine the equal-division coalitions as in the core? The answer is that the restriction to sequential proposals, when coupled with the slight discounting, rules out two competing offers being simultaneously considered, thereby ruling out most undercutting and overbidding.

This model involves time discounting, a restriction to sequential offers and an infinite horizon. None are factors in our experiment. Also, a previous experiment on a finite horizon, random proposer variant of this extensive form (Bolton and Chatterjee 1996, see

---

[19] Chatterjee and Dutta (1998) study public and private offers in a game with two heterogeneous buyers and two identical sellers, with offers from each side of the market being simultaneously made, as are responses from the side receiving offers. The public offers game has an equilibrium in the core, which is destroyed by heterogeneity on both sides of the market. Under private offers, the (public perfect, mixed) equilibrium involves a positive probability of delay and is not in the core.

[20] 33% in public; 42% in unconstrained; 30% in C-controls; 65% in S-controls; and 31% in T-controls.

[21] It was the post hoc examination of this model that led us to look for competitive frictions in our data.

Okada 1996 for the theoretical analysis of the infinite horizon, random proposers model) found little evidence for subgame perfect equilibrium play. Nevertheless, the model demonstrates that it is possible to construct a noncooperative game with a competitive friction (not entirely unlike the sort we have been considering), and with results bearing more than a passing resemblance to what we observe.

6. Summary

If we had confined ourselves to examining any single communication configuration, we might have drawn substantially different conclusions than we have from examining a more comprehensive set. Had we confined ourselves to comparing unconstrained communication with one in which S-controls communication, for example, we would have found that communication constraint has little influence on the negotiation. Systematically varying the communication configurations, however, reveals a pattern of clear strategic influence.

To summarize the important points of our study

• *Constraining communication had an effect on the pattern of coalition formation in every case save for giving the strongest bargainer control of communication.* Giving communication control to the two weaker bargainers benefited these bargainers, and also led to more grand coalitions. Restricting to public communication led to overall more competitive-looking payoff allocations within coalitions. Eliminating the connected communication path within a coalition eliminated that coalition from the settlements.

• *A model that can explain the equal-division coalitions will also likely explain many of the other phenomena we observe.* Nearly equal-division allocations are associated with the coalition that has the highest per capita value. Once we accept the nature of these coalitions, many of the communication regularities follow rather naturally from an examination of the competitive pressures in the environment.

• *A propensity to make private offers to prospective coalition partners appears to be a key strategic factor.* When feasible, private offers of two-person coalitions are the rule, and the one treatment where they

are not feasible, public, exhibits a distinct pattern of settlement.

Our findings suggest that parties involved in multilateral bargaining be cognizant of the opportunities and constraints presented by the communication structure. Parties with weaker alternatives would benefit from a more constrained structure, especially if they can be the conduit of communication, while those endowed with stronger alternatives would do well to work within a more public communication structure that promotes competitive bidding. Real estate agents intuitively understand that their power (and, hence, their payoffs) relies on their control over the information passed between potential buyers and sellers. Similarly, Pakistan's parlaying of communication between the United States and China in the 1970s was the key to its advantage.

The payoffs chosen for study here had three important properties: (1) There was a clear ranking of strength across the players, (2) a two-person coalition had the greatest per capita value, and (3) there was a unique allocation in the unconstrained core. It would be informative to repeat the experiment varying each of these properties. Additional communication issues arise with four or more bargainers, and these remain to be investigated. With four or more, there could be players who want to wait for some other coalition to form so as to improve their own competitive position with respect to those players not in the first coalition.

Our findings point to the importance of analyzing the bargaining process, as well as the outcome regularities associated with communication. Even in free-form games, we find that parties do not bargain in completely unconstrained fashion—their moves are constrained by seemingly shared beliefs about acceptable procedures, a preference for private proposals, and a tendency, at least in some cases, to push agreement to deadline. This behavior finds parallel in the field: In some of the merger examples we cited in the introduction, bargaining parties could freely communicate with all those involved, but chose not to do so. The importance of deadlines in field bargaining is well documented. Further study of the process of coalition bargaining could shed valuable light on these complex but important negotiations.

Acknowledgments

The authors gratefully acknowledge the financial support of Fuji-Xerox Ltd. through a grant to the Penn State Center for Research in Conflict and Negotiation, and the support of the Harvard Business School. Gary Bolton gratefully acknowledges the support of the National Science Foundation. Kalyan Chatterjee thanks the American Philosophical Society for a sabbatical fellowship for the period in which this draft was prepared. The authors thank Michikazu Aoi for his input in several of our initial discussions. They are grateful to Vijay Krishan and David Laibson for suggestions, and seminar audiences in Hong Kong, Lund, New York, and London for comments. The authors also thank Julia Morgan for assistance in gathering the data and James Evans for assisting in the data analysis.

Appendix A: Written Instructions to Subjects

General. Please read the instructions carefully. If at any time you have questions or problems, raise your hand and the monitor will be happy to assist you. From now until the end of the session, unauthorized communication of any nature with other participants is prohibited.

During the session, you will engage in a series of negotiations, carried out over e-mail with other participants. Each negotiation gives you an opportunity to earn cash.

Description of the Negotiation. Each negotiation involves representatives from three cement-making companies: the Scandinavian Cement Company (*SC*), the Cement Corporation (*CC*), and the *Thor* Cement Company (*Thor*). The three companies are contemplating a formal merger. Each firm would bring value to the merger greater than its own individual profit, because of the synergy that would be realized. But how much extra value depends on the mergers that are formed. The following schedule shows the total profit value of all possible mergers, in a fictional currency called thalers (how thalers translate into dollars is explained below):

Merging Parties	Total Profit of Merger (in Thalers)
SC, *CC*, and *Thor*	100
SC and *CC*	90
SC and *Thor*	70
CC and *Thor*	40

Only one merger is allowed per negotiation. Before entering into a merger, representatives from the merging companies must agree on how to divide the profits. The goal of each representative is to get the most profits possible for his or her own company.

You have the role of a company representative. The company you represent is indicated on the top, right-hand corner of this sheet.

Each negotiation lasts, at most, 8 minutes. During this time, company representatives may bargain with one another. All negotiations are conducted by computer e-mail. Instructions on how to operate the e-mail will be provided after everyone has finished reading the instructions. The rules of communication are as follows:

> *SC* may send a message to either *CC* or *Thor* or to both.
> *CC* may send a message to either *SC* or *Thor* or to both.
> *Thor* may send a message to either *SC* or *CC* or to both.

OR

> *Thor* may send a message to either *SC* or *CC* or to both.
> *CC* may send a message to *Thor* but not to *SC*.
> *SC* may send a message to *Thor* but not to *CC*.

The only way for SC or CC to get a message to one another is to send it through Thor.

Please note: Under no circumstances are you allowed to identify yourself by your real name.

Once an agreement is reached, the merging parties must complete a contract stipulating the terms of the agreement. In order for the contract to be valid, a contract form must be completed by each merging party by the time the monitor announces the end of the 8-minute period. A merger is valid only if each merging party has entered the same information about partners and profits onto the form. In addition, any contract has total profits exceeding the limits set by the above schedule will be considered invalid. All companies not in a contract at the end of the negotiation receive zero (0) profit for that negotiation.

Contract Form. At the conclusion of each negotiation, all of the parties included in the agreed-upon merger must fill out a contract form for each negotiation. There are a number of these forms attached. Fill in how much each party to the agreement receives, note what role you are playing, state your e-mail ID (e.g., Mock 4), and sign your name. Each person will have an identical form. We will pay you what the contract states you have earned *only* if the forms from all the parties in the agreement match.

Negotiation Record. A blank "History" form is provided in your folder. At the conclusion of each negotiation, please fill out this form. The completed form provides you with a history of your past negotiations, and you may reference it at any time during the session.

Grouping Procedure. You will change bargaining partners for each negotiation. You will never negotiate with the same person more than once. Partner identities are confidential and will not be revealed at the end of the session. Please do not identify yourself to the other party during the negotiation.

Earning Money for Yourself During the Negotiations. You will negotiate more than once. You will actually be paid for just one negotiation. The payoff negotiation will be selected by a lottery after all the negotiations have been completed. Each negotiation has an equal chance of being selected, so it is in your interest to make as much profit as you can in each and every one. Each thaler you earn is worth \$.50. For example, if the second negotiation were selected in the lottery and you earned 40 thalers in that negotiation, you would be paid $20. Immediately upon conclusion of the session, you will be paid your earnings in cash.

Appendix B: Rotation of Bargaining Partners

With 15 subjects, it is possible to have each subject play 5 games in the same role, and never assign any subject to play with any other subject more than once. The algorithm for assigning bargaining partners begins with the chart below. Each row number corresponds to a subject assigned the roles of *SC*, and each column number to someone assigned *CC*, and the entries in the chart to those assigned *Thor*.

Subject	2	5	8	11	14
1	3	6	9	12	15
4	6	9	12	15	3
7	9	12	15	3	6
10	12	15	3	6	9
13	15	3	6	9	12

The first set of games, along the main diagonal, would be 1 2 3, 4 5 9, 7 8 15, 10 11 6, 13 14 12. The next set of games would be 4 2 6, 7 5 12, 10 8 3, 13 11 9, 1 14 15; followed by 7 2 9, 10 5 15, 13 8 6, 1 11 12, 4 14 3; 10 2 12, 13 5 3, 1 8 9, 4 11 15, 7 14 6; 13 2 15, 1 5 6, 4 8 12, 7 11 3, 10 14 9. Thus, subject 1 would play with subjects 2, 3; 14, 15; 11, 12; 8, 9; 5, 6. Therefore, 1 would play with all subjects with the *CC* role (2, 5, 8, 11, 14) as well as all subjects with the role of *Thor* (3, 6, 9, 12, 15). Subject 5 (*CC*) would play the following sequence of opponents: 4, 9; 7, 12; 10, 15; 13, 3; 1, 6. The other subjects would be assigned in the same way.

References

Aumann, R. J., R. B. Myerson 1988. Endogenous formation of links between players and of coalitions: An application of the Shapley value. A. E. Roth, ed. *The Shapley Value: Essays in Honor of Lloyd S. Shapley.* Cambridge University Press, Cambridge, U.K., 176–191.

Binmore, K., J. Swierzbinski, S. Hsu, C. Proulx. 1993. Focal points and bargaining. *Internat. J. Game Theory* **22** 381–409.

Bolton, G., K. Chatterjee. 1996. Coalition formation, communication and coordination: An exploratory experiment. R. Zeckhauser, R. Keeney, J. Sebenius, eds. *Wise Choices: Games, Decisions, and Negotiations.* Harvard Business School Press, Boston, MA, 253–271.

Borm, P., G. Owen, S. Tijs. 1992. On the position value for communication situations. *SIAM J. Discrete Math.* **5** 305–320.

Bottom, W. P., J. Holloway, S. McClurg, G. J. Miller. 2000. Negotiat

ing coalitions: Risk, quota shaving, and learning to bargain. *J. Conflict Resolution* **44** 147–169.

Chatterjee, K., B. Dutta. 1998. Rubinstein auctions: On competition for bargaining partners. *Games Econom. Behavior* **23** 119–145.

——, ——, D. Ray, K. Sengupta. 1993. A noncooperative theory of coalitional bargaining. *Rev. Econom. Stud.* **60** 463–477.

Cialdini, R. B. 1993. *Influence: The Psychology of Persuasion.* William Morrow, New York.

Diermeier, D., R. Morton. 2000. Proportionality versus perfectness: Experiments in majoritarian bargaining. Working paper, Northwestern University, Evanston, IL.

Murnighan, J. K., A. E. Roth. 1977. The effects of communication and information availability in an experimental study of a three-person game. *Management Sci.* **23** 1336–1348.

Myerson, R. B. 1977. Graphs and cooperation in games. *Math. Oper. Res.* **2** 225–229.

Okada, A. 1996. A noncooperative coalitional bargaining game with random proposers. *Games Econom. Behavior* **16** 97–108.

——, A. Riedl. 2001. Reciprocity, inefficiency and social exclusion: Experimental evidence. Working paper, Tinbergen Institute, Amsterdam, The Netherlands.

Panorama. 1999. *Mining Magazine* **181**(November) 284.

Pilkington, A. 1996. *Transforming Rover: Renewal Against the Odds— 1991–1994.* Bristol Academic Press, Bristol, U.K.

Raiffa, H. 1982. *The Art and Science of Negotiation.* Harvard University Press, Cambridge, MA.

Rapoport, A., J. P. Kahan. 1976. When three is not always two against one: Coalitions in experimental three-person cooperative games. *J. Experiment. Soc. Psych.* **12** 253–273.

Roth, A. E. 1995. Bargaining Experiments. J. Kagel, A. E. Roth, eds. *Handbook of Experimental Economics.* Princeton University Press, Princeton, NJ, 253–348.

Sebenius, J. K. 1984. *Negotiating the Law of the Sea.* Harvard University Press, Cambridge, MA.

Selten, R. 1987. Equity and coalition bargaining in experimental three-person games. A. E. Roth, ed. *Laboratory Experimentation in Economics: Six Points of View.* Cambridge University Press, Cambridge, U.K., 42–98.

Valley, K. L., S. Blount White, M. A. Neale, M. H. Bazerman. 1992. Agents as information brokers: The effects of information disclosure on negotiated outcomes. *Organ. Behavior Human Decision Processes* **51** 220–236.

Van Huyck, J., R. Battalio, F. W. Rankin. 1997. On the origin of convention: Evidence from coordination games. *Econom. J.* **107** 576–596.

——, ——, S. Mathur, P. Van Huyck. 1995. On the origin of convention: Evidence from symmetric bargaining games. *Internat. J. Game Theory* **24** 187–212.

Von Neumann, J., O. Morgenstern. 1953. *Theory of Games and Economic Behavior,* 2nd ed. Princeton University Press, Princeton, NJ.

Accepted by Martin Weber; received August 20, 2001. This paper was with the authors 4 months for 1 revision.

Quarterly Journal of Political Science, 2011, 6: 1–53

Pre-electoral Coalitions and Post-election Bargaining*

Siddhartha Bandyopadhyay[1], Kalyan Chatterjee[2] and Tomas Sjöström[3]

[1] Department of Economics, University of Birmingham, UK;
s.bandyopadhyay@bham.ac.uk
[2] Department of Economics, The Pennsylvania State University, USA;
kchatterjee@psu.edu
[3] Department of Economics, Rutgers, The State University of New Jersey,
USA; tsjostrom@econ.rutgers.edu

ABSTRACT

We study a game-theoretic model where three political parties (left,
median and right) can form coalitions both before and after the
election. Before the election, coalitions can commit to a seat-sharing
arrangement, but not to a policy platform or a division of rents from
office; coalition members are free to break up and join other coalitions after the election. Equilibrium pre-electoral coalitions are not
necessarily made up of the most ideologically similar parties, and they
form under proportional representation as well as plurality rule. They
form not only to avoid splitting the vote, but also because seat-sharing
arrangements will influence the post-election bargaining and coalition

* We thank Facundo Albornoz, Sophie Bade, Ralph Bailey, Somdutta Basu, Myeonghwan
  Cho, Jayasri Dutta, John Fender, Indridi Indridason, Saptarshi Ghosh, Bryan McCan-
  non, and especially Sona Golder for valuable comments. We also thank two anonymous
  referees and the editors for many helpful comments. Chatterjee wishes to thank the
  Human Capital Foundation (www.hcfoundation.ru), and especially Andrey P. Vavilov,
  for support to The Pennsylvania State University's Department of Economics.

MS submitted 26 May 2011; final version received 3 July 2011
ISSN 1554-0626; DOI 10.1561/100.00010043

formation. The median party's share of the surplus in a two-party government is large if ideology is not very important, or if its ideological position is not very distant from the third (outside) party, so that it has a credible threat to switch coalition partners. On the other hand, if ideology is very important, and if the right and left parties are ideologically distant from each other so each is willing to give up a lot to prevent the other from joining a governing coalition, then the equilibrium outcome may be that the median party forms a one-party government.

In parliamentary democracies, coalition governments are common, and single party majority governments are relatively rare. A study of 313 elections in 11 European democracies between 1945 and 1997 found that only 20 elections returned a single party with more than half of all seats in parliament (Gallagher *et al.* 1995, Diermeier and Merlo 2004, Strom *et al.* 2008: Chapter 1). But coalitions can form both ex ante (before elections) and ex post (after elections). There is a well developed literature on post-electoral coalition formation, but as noted by Powell (2000, p. 247), the literature on pre-electoral coalitions is quite small.

Recent empirical work suggests that pre-electoral coalitions are important. In a study of 364 elections in 23 advanced parliamentary democracies between 1946 and 2002, Golder (2006a, b) found 240 instances of pre-electoral agreements.[1] Carroll and Cox (2007) found that of the 144 parties participating in majority governments in their cross sectional data, 38 (26%) had engaged in public pre-election cooperation. Pre-electoral agreements are common in diverse countries such as France, South Korea and India. Debus (2009) offers empirical evidence that pre-electoral alliances have an impact on government formation. In this paper, we investigate theoretically how different electoral systems and post-election bargaining protocols influence the process of pre-electoral coalition formation.

It has long been recognized that under plurality voting (PV), like-minded parties who compete against each other in the same electoral districts risk splitting the vote. In 1903, the U.K. Labour party and Liberal Democrats

---

[1]   Martin and Stevenson (2001) consider only a single data set and do not report the percentage of elections with pre-electoral coalitions (which is not the focus of their paper), but Golder (2006a, b) calculates that the percentage in their data set is 19%. She argues, however, that this is an underestimate.

Pre-electoral Coalitions and Post-election Bargaining 3

formed the first Lib-Lab pact, in which they agreed not to compete against each other for 50 seats in parliament (Pugh, 2002, p. 117). Various forms of Lib-Lab arrangements persisted, mainly in local elections in Scotland and Wales, though attempts at national seat-sharing agreements have also been made.[2] In India, which also has a PV electoral system, pre-election coalitions became widespread following the 1977 election, when the Indian National Congress lost its hold on power.[3] Typically, these pre-electoral coalitions do not commit to form a coalition government, which will implement a specified set of policies. Instead, the main issue over which pre-electoral alliances in India bargain is which party will contest which seat.

Under a system of proportional representation (PR) with national lists, as in Israel, each list gets a number of seats in parliament proportional to its vote share. If two parties stand on a joint list, and if each voter who supports either party votes for the joint list, then the joint list will get the same number of seats in parliament as the two parties would get by standing on separate lists. Thus, in this system, the problem of splitting the vote is moot, seemingly eliminating the rationale for ex ante agreements. But in reality, ex ante coalitions occur even with proportional representation. For example, 87% of the elections in Israel (which comes closest to pure PR) analyzed in Golder's data set had at least one pre-electoral alliance. Similarly, joint lists have been seen in Greece, Portugal and (to a lesser extent) the Netherlands (Golder 2006a, b).[4] Our theoretical model investigates the possible motives for such pre-electoral agreements.

Most real-world PR systems are characterized by a combination of national list choice and district level elections. However, in order to understand the motives for ex ante coalitions, we will study a pure system of strictly proportional representation with national lists where the problem of splitting the vote is absent. Under this voting system, an ex ante coalition is simply an agreement to contest the election as a single national list. The ordering of candidates on the list will determine the parties' number of seats in parliament. We also study a second voting system, plurality voting (PV), where

---

[2] In contrast, the 2010 Conservative–Liberal Democrat coalition resulted from post-electoral negotiations.

[3] See http://www.electionresults.in/history-political-parties.html for a brief history of how the Indian National Congress lost its hold on power.

[4] Empirical estimates of the frequency of coalitions depend on the exact definition which is used. In our model, pre-electoral coalitions do not commit to forming a coalition government, or to any policy platform. Therefore, as a motivation for our work, the most permissive definition of pre-electoral coalition (and hence biggest number) seems the most relevant.

the electorate is divided into districts and each district elects a member of parliament. Under this voting system, an ex ante coalition is an agreement not to compete in certain districts. This may not be a complete seat-sharing arrangement; there may be some seats in which both parties run for office.[5]

In our model there are three parties, L, M and R, with M ideologically closer to L than to R.[6] The parties care about ideology, rents from office and seats in parliament. If ideologically distant parties form a coalition government, they may experience costs of ideological compromises. Therefore, an MR coalition government (consisting of the M and R parties) generates a smaller surplus than an LM government (consisting of L and M). For simplicity, we assume the L and R parties are so far apart ideologically that an LR coalition cannot generate any surplus. If a party is outside the government, it may suffer a negative externality from a government to which it is ideologically opposed. Ex ante coalitions determine the seat shares of the coalition partners, but they are free to split up after the election. If no party obtains a majority of the seats in parliament, then post-election bargaining determines which government forms, and how the rents from office are allocated.

The timing is as follows. First, pre-electoral coalitions form. Then voting occurs (under PR or PV). Finally, post-election government formation takes place. This stark model explores the pure incentives for coalition formation, emphasizing the role of ex post bargaining and coalition formation, while abstracting from issues such as increasing returns to scale in campaign effort for ex ante coalitions.

We consider two canonical ex post bargaining protocols. The *random recognition protocol* specifies that, in each period of bargaining, each party is recognized to propose a coalition with probability proportional to its number of seats in parliament. Similar protocols have been analyzed by Baron and Ferejohn (1989) and others. The *ASB protocol* (after Austen-Smith and Banks 1988) is deterministic: the largest party is recognized first, followed

---

[5] For example, in 2001 in Assam (one of the states in India), the BJP and AGP parties agreed that the BJP would put up candidates for 44 seats, but 10 of these would be contested by both parties in so-called friendly contests (http://news.indiamart.com/news-analysis/assembly-polls-congr-6008.html).

[6] This simplification is made in order to pinpoint the tradeoffs and incentives for coalition formation. Many countries do have only three major parties, e.g., the U.K. (Labour, Conservative, Liberal Democrat) and Israel (Likud, Kadima, Labour).

by the second largest and so on.[7] We characterize the stationary subgame perfect equilibria of the infinite horizon ex post bargaining games corresponding to the two protocols. With random recognition, the MR coalition government never forms. Equilibrium surplus shares within the LM coalition government are proportional to seat shares (Gamson's Law) if M and R are ideologically very distant. However, M's surplus share is bounded below by the surplus an MR government would generate, so Gamson's Law is violated if M and R are ideologically fairly close. With the ASB protocol, for some orders of recognition, the MR government forms if M and R are ideologically fairly close.

There are three motives for ex ante coalition formation in our model: (a) to influence which government will form ex post; (b) to manipulate the bargaining power within the government; and (c) with plurality voting only, for similar parties to avoid splitting the vote. We emphasize (a) and (b), as (c) is well known (Golder, 2006a, b; Blais and Indridason, 2007). One way for motive (a) to come about is via an ex ante agreement which produces such a large vote share for M that it becomes a majority party. The junior ex ante coalition partner, say R, benefits from this seemingly one-sided agreement because it blocks its ideological opponent L from joining a coalition government. With ASB bargaining ex post, there is another way for (a) to happen: whether MR or LM forms ex post may depend on the order of recognition, and this can be influenced by transferring seats from one party to another. Motive (b) can come about via an ex ante agreement that transfers enough seats to change the ex post distribution of surplus, via the ex post bargaining protocol, without actually changing the governing coalition. Because of (a) and (b), ex ante agreements may be viable under PR. Also, because of these motives, under PV ideologically different parties (M and R) unconcerned about splitting the vote may still find a viable ex ante agreement. Thus, one of our main conclusions is that, in theory at least, ex ante coalitions are by no means motivated solely by the problem of splitting the vote.

To close the model, we assume *ex ante* bargaining follows a random recognition protocol with a deadline given by the election. Party M is the essential party which must be part of any coalition. The more rounds of bargaining ex ante, the better M can exploit this position, so its equilibrium payoff is

---

[7] Diermeier and Merlo (2004) argue that there is greater empirical support for the random recognition protocol, yet this is questioned by Laver *et al.* (2010). Rather than take sides we consider both protocols.

increasing in the number of rounds. If ideology is not very important, or if M's ideological position is not very distant from that of R, then M has a credible outside option and will get a large share of the surplus in a coalition with L. On the other hand, ideological polarization means L and R are keen on excluding each other from the government. If these parties care much more about ideology than about seats in parliament then M benefits from the polarization; indeed, if there are many rounds of bargaining ex ante then M will be able to form a majority government and take all the rents from office. This and other predictions, suggesting some ideas for empirical work, are discussed in the conclusion.

There is a large game theoretic literature on bargaining and coalition formation (e.g., Chatterjee *et al.* 1993; Okada 1996, 2007; Eraslan and Merlo, 2002). Nearly all coalition formation papers which allow externalities assume symmetric players (Ray, 2008). Our analysis of *ex post* legislative bargaining may be of independent interest, as we assume heterogeneous players and externalities, and we characterize and compare the stationary subgame perfect equilibria for different extensive forms. Our random recognition protocol does not require the game to be superadditive, and non-degenerate mixed strategies are necessarily used in equilibrium. From the political economy angle, we derive endogenous shares of the surplus based on the proportion of seats in the legislature. Our analysis of a finite horizon *ex ante* coalition formation game might also be of independent theoretical interest.

Starting with Riker (1962), a large literature in political science discusses coalition formation in legislatures (e.g., Laver and Schofield, 1990; Roemer, 2001; Bandyopadhyay and Chatterjee, 2006). Riker considered the allocation of rents from office, and Axelrod (1970) added ideological motives. Austen-Smith and Banks (1988) provide a formal game-theoretic model of how the nature of coalitions (ex post) influence voting. Diermeier and Merlo (2000) and Baron and Diermeier (2001) study post-election coalitional diversity. Indridason (2003, 2005) empirically studies what factors affect the size and connectedness of coalitions and Bandyopadhyay and Oak (2004, 2008) develop a theoretical model.

Golder (2006b) not only focuses on the empirical study of pre-electoral coalitions, but also presents a simple theoretical model. Our model differs from Golder's in several ways. First, we model political competition explicitly: parties have a choice of coalition partners. In Golder's model, the identity of the coalition partner is not a choice variable (the choice is only whether to accept this partner or not). Second, we explicitly model the voting process. Third, Golder assumes pre-electoral coalitions make binding commitments on policy and rents from office. In our model, pre-electoral coalitions agree on seat-sharing arrangements or joint lists, but can make no other commitments (on future policies, surplus-sharing or government formation).

While there is no agreement in the literature about what parties can commit to, the perfect commitment assumption of the Downsian model (Downs, 1957) is often considered unrealistic. Citizen candidate models assume no commitment (Osborne and Slivinski, 1996; Besley and Coate, 1997). In dynamic models of legislative bargaining, current policies can influence future outcomes, which allows for a limited kind of commitment. Baron *et al.* (2011) explicitly model the dynamics of the electoral process. They rule out commitment to future actions such as government formation or policy choice. However, in their model, today's policy becomes tomorrow's status quo, and so will influence future outcomes. In our model, a different kind of partial commitment, namely ex ante seat-sharing arrangements or joint lists, is used to manipulate ex post outcomes.

Baron *et al.* (2011) assume voters behave strategically, taking into account that today's policy will become tomorrow's status quo. Parties strategically choose today's policy, taking into account that this will influence not only future legislative bargaining and policy outcomes, but also the behavior of the voters in the next election. In our basic model voting is sincere, and pre-electoral agreements amount to directly manipulating the election outcome, which in turn will determine government formation and legislative bargaining power. This suggests that, if the voters' interests are aligned with their party leaders, strategic coordination of voter behavior could replicate the outcome of pre-electoral seat sharing arrangements, but we show that this is not the case.

In our model, seat sharing arrangements (or joint lists) are negotiated ex ante. Ex post, the parties negotiate a government and share the rents from office the government generates. These rents can be identified with ministries allocated among the coalition partners (Laver and Schofield, 1990; Laver and Shepsle, 1996). Seats in parliament are non-transferable ex post. Indeed, there are explicit laws or strong norms against seat-selling (or even floor-crossing) in almost all constitutional democracies. We also rule out the possibility that seats generate transferable benefits that can be traded for policy concessions or used to induce a majority party to share power. Indeed, in most U.K. elections, for example, the party with a majority has formed a single party government. This was also the case in India until the 1980s, when a single party used to get a majority. Plainly, other parties did not have anything to offer the majority party that would make power sharing worthwhile. Indeed most empirical analyses of coalition formation implicitly assume this, by excluding all cases with a majority party from the analysis (e.g., Martin and Stevenson, 2001). Other transfers, such as direct cash payments, are also ruled out. In short, following Austen-Smith and Banks (1988), we assume the rents from office generated by the government is the *only* benefit that can be transferred *ex post*. In our model, if a party wins a majority then it will always form a single party government, implement its favorite policy and take all the rents from office generated by the government, because it has nothing to gain from power sharing.[8] This seems consistent with the empirical evidence cited above.

There is evidence that pre-electoral coalitions influence post-election government formation (Debus, 2009; Golder, 2006b). But pre-electoral alliances often break up, with former coalition partners not cooperating in forming a government, suggesting less than perfect commitment. For example, the Janata Party, a merger of various groups opposed to the Congress, won the national election in India in 1977. After a few years, the Janata Party split into its components, and these have since formed a number of pre-electoral coalitions. These coalitions are clearly not mergers; the parties consider themselves free to join different post-electoral coalitions.[9]

With perfect commitment, ex ante agreements would be akin to form a new party. Dhillon (2005) surveys the party formation literature. Morelli (2004) assumed new parties form by mergers involving binding commitments on policy and ex post cooperation.[10] In our model, a pre-electoral coalition does not signify a merger where the parties give up their separate identities.

---

[8]   Notice that the outcome may not maximize the total payoff of the three parties, because of the externalities.

[9]   For example, *The Hindu* newspaper of May 15, 2009 reported that Nitish Kumar of the Janata Dal (United) party, a member of the pre-electoral alliance National Democratic Alliance, stated his conditions for supporting any coalition government, including one *not* formed by the National Democratic Alliance. Several members of the pre-electoral alliance Third Front also declared themselves ready to switch to other groupings after the election.

[10] Other works on party formation includes Roemer (2001), Jackson and Moselle (2002), Snyder and Ting (2002), Levy (2004) and Osborne and Tourky (2002). Pech (2010) analyzes a scenario where binding pre-electoral agreements are possible, parties are policy motivated and voting is sincere. There is no coalitional bargaining and policy of a coalition government is simply a lottery over the ideal positions of its constituent parties. He shows that there is a unique *stable set* where the median party runs uncontested.

Instead, the parties remain independent and (as long as no party has its own majority) must bargain ex post to form a coalition government. In addition, unlike in Morelli (2004), our parties get utility not only from seats in parliament, but also from joining the government, and even from blocking ideologically distant parties from joining. The issue of maintaining separate identities versus mergers is also analyzed by Persson *et al.* (2007). Their parties (unlike ours) are opportunistic and represent specific constituencies and not ideological positions, and their focus is on comparing government spending under single party versus coalition governments.

The rest of the paper proceeds as follows. After presenting the model, we characterize *post-election* bargaining equilibrium under the two protocols. We then turn to the issue of *ex ante* coalitions. Before concluding, we briefly discuss strategic voter coordination.

The Model

Parties, Voters and Preferences

There are three parties arranged from left to right: L, M and R. There are three kinds of voters: L-supporters, M-supporters and R-supporters. Voter preferences are such that L-supporters rank party L first, party M second and party R last, and R-supporters rank R first, M second and L last. Without loss of generality, we assume the M party is ideologically closer to the L party, so the M-supporters rank M first, L second and R last. Let $v(P)$ denote the fraction of all voters who support party $P \in \{L, M, R\}$. To avoid trivialities we assume $0 < v(P) < 1/2$ for each $P \in \{L, M, R\}$. For convenience, we normalize the total number of seats in parliament equal to 1. Party P's share of the seats in parliament is denoted by $n(P)$, where $n(L) + n(M) + n(R) = 1$.

With a continuum of voters, no single voter can change the outcome of the election. If the P-supporter derives some utility from voting according to his ideology as well as from the ultimate outcome, then it is clearly optimal to vote for party P whenever possible. We assume such sincere voting for most of this article, following Fong (2006) and other models of legislative bargaining.[11] If party P does not have a candidate in a district, the

---

[11] Empirical studies suggest this captures the behavior of most voters (Degan and Merlo, 2007; Brunell and Grofman, 2009).

P-supporters vote for their second most preferred party, for the same reason. However, if positive measures of voters are allowed to coordinate their votes, they might change the outcome of the election, and we discuss such strategic vote coordination in a separate section.

Each party is considered an individual player who derives utility from seats in parliament. Let α denote the value of a seat, which is the same for all parties. In addition, if a party is a member of government, it enjoys a share of the surplus generated by the government, the rents from office.[12] Parties also care about policy, in two respects: (i) if they join a coalition government then they face a compromise cost which is lower if the partners are ideologically closer; and (ii) if they are not in government then they suffer a cost from the policy implemented by the party (or parties) in government, the cost being lower if the government is ideologically closer to them.

A one-party government generates a surplus S. Presumably, the ruling one-party government will simply implement its ideal policy. In contrast, a coalition government must make a costly compromise between the policy preferences of the coalition partners. Thus, a two-party coalition government, consisting of parties P and P', generates a surplus $S(PP') = S - c(PP')$, where $c(PP') > 0$ is its compromise cost. The compromise cost is greater, the more ideologically distant are the two parties. Thus, $c(MR) > c(LM)$.[13] To avoid a proliferation of special cases, we assume the most ideologically distant parties L and R together cannot generate any surplus, so the LR coalition government will never form. In summary, we assume the various kinds of governments generate the following net surpluses, also

---

[12]   The rents from office generated by the government are, in contrast to the value of a seat in parliament, transferable. By "rents" or "surplus" we mean these transferable rents generated by the government, not by the non-transferable value of a seat.

[13]   For example, consider a standard Hotelling model with policy space $[0, 1]$. Party P's ideal point is $y(P) \in [0, 1]$, where $y(L) < y(M) < y(R)$. A one-party government would simply implement its own ideal policy. If instead parties L and M form a coalition government, they willl implement some compromise policy $y(LM)$. The compromise would satisfy $y(L) < y(LM) < y(M)$. (Its precise location would depend on the relative powers of parties L and M as determined by, for example, their seat shares.) The "travel cost" for a coalition partner is the distance between the implemented policy and its ideal point. The compromise cost for the LM government is its total travel cost,

$$c(LM) = |y(LM) - y(L)| + |y(M) - y(LM)|$$
$$= y(M) - y(L) > 0.$$

Similarly, the compromise cost for an MR government would be $c(MR) = y(R) - y(M)$. Since L and M are closer ideologically, their ideal points are closer, i.e., $y(R) - y(M) > y(M) - y(L)$. Thus, $c(MR) > c(LM)$.

Pre-electoral Coalitions and Post-election Bargaining 11

referred to as the *rents from office*:

$$0 = S(LR) < S(MR) < S(LM) < S.$$

If party P is part of the government, then let $s(P) \geq 0$ denote its share of the surplus. For a one-party government there is no compromise cost so $s(P) = S$. In a two-party government, the coalition partners P and P' must agree on some (non-negative) shares $s(P)$ and $s(P')$, subject to the constraint

$$s(P) + s(P') = S(PP').$$

A government may impose negative externalities on outsiders (say, by implementing policies they don't like). Formally, if party P is *not* a member of government, it suffers a cost $x_P(P'P'')$ if the other two parties P' and P'' form a coalition government, and $x_P(P')$ if party P' forms a one party government. We assume

$$0 \leq x_P(M) < x_P(MP')$$

for $P, P' \neq M$. That is, each party $P \in \{L, R\}$ prefers a one-party M government to a coalition government where M governs together with the other party $P' \neq P$.[14]

We can now summarize the payoff structure. If party P is part of the government, then its payoff is $s(P) + \alpha n(P)$. If party P is not part of the government, then its payoff is $\alpha n(P) - x_P$, where $x_P = x_P(P')$ if party P' forms a one-party government, and $x_P = x_P(P'P'')$ if P' and P'' form a coalition government. For example, if M and R form a coalition government, then party L's payoff is $\alpha n(L) - x_L(MR)$.

---

[14] Consider the Hotelling framework of the previous footnote. If parties L and M form the government and implement policy $y(LM)$, party R suffers a travel cost

$$x_R(LM) = |y(R) - y(LM)|.$$

In contrast, if M forms a one-party government and implements $y(M)$, party R's travel cost is

$$x_R(M) = |y(R) - y(M)|.$$

Since $y(LM) < y(M)$, the spatial framework (with linear transportation cost) implies

$$0 < c(MR) = x_R(M) < x_R(LM).$$

Thus, party R's ideological loss from having M's policy implemented is lower than the loss from the compromise policy of the LM coalition, which would be further away from R's ideal.

Elections

We consider two kinds of voting systems: proportional representation (PR) and plurality voting (PV). Proportional representation is a national election in which lists compete against each other. If all parties run for election on separate lists, then proportional representation implies $n(P) = v(P)$ for each $P \in \{L, M, R\}$.

To describe the outcome of plurality voting, we assume the electorate is divided into a large number of ex ante identical districts.[15] We assume that the overall results of the election can be predicted with certainty ex ante, and this can be justified because the number of districts is very large and there are no aggregate shocks. However, since the districts are ex ante identical, but experience idiosyncratic shocks to the election results, it is not possible to predict which particular districts will be won by which party. If all parties run for election in every district, then party $P \in \{L, M, R\}$ wins a *plurality* in a fraction $w(P)$ of all districts, and a *majority* in a fraction $z(P)$ of all districts, where $0 < z(P) < w(P) < 1/2$. Thus, under PV, if all parties run in each district then $n(P) = w(P)$ for each $P \in \{L, M, R\}$. Note that $v(P) \neq w(P)$ in general.[16] Also, to simplify and eliminate some less interesting cases, we assume it is not too likely a party wins a majority in any district. Specifically, we assume

$$z(P) < \min\{w(L), w(M), w(R)\}$$

for each $P \in \{L, M, R\}$.

---

[15] Heterogeneity could be introduced by considering districts in the set $[0, 1]$, with an interval $[0, z(L)]$ representing safe seats for party L. The interval $[z(L), w(L)]$ would represent seats that L would win in a three-way contest but lose if it were running solely against M. The remaining sub-interval would be partitioned in a similar way. This generalization is left for future work.

[16] For example, suppose the districts are ex ante symmetrical, but when elections occur there is a random variable x_i for district i that takes one of three values (L, M or R), each value occurring in a third of the districts. If $x_i = R$, then L has support of 30% of the voters in district i, M has 20%, and R has 50%. If $x_i = L$ (resp. $x_i = M$) the numbers are 60% for L, 30% for M, 10% for R (resp. 30%, 40%, 30%).

	0.33	0.33	0.33
L	0.3	0.6	0.3
M	0.2	0.3	0.4
R	0.5	0.1	0.3

Here $w(P) = \frac{1}{3}$ for each party. However, the nationwide vote share is $v(L) = 0.4$, $v(M) = v(R) = 0.3$.

If party P gets more than half of all seats in parliament, i.e., $n(P) \geq 1/2$, then P forms a one-party government and the game ends. Otherwise, i.e., if $n(P) < 1/2$ for all P, there will be post-election bargaining. Notice that if $n(P)$ is exactly 50%, we consider that party P has a majority. This will simplify the discussion of ex ante bargaining, but does not have any real significance.

Post-Election Bargaining

If no party has a majority of the seats in parliament, then two parties P and P' can form a coalition government. Within the governing coalition, utility can be transferred (only) by allocating the surplus $S(PP')$ the government generates. A proposal to form a PP' coalition government specifies how these rents from office are to be shared: P gets $s(P) \geq 0$ and P' gets $s(P') \geq 0$, where $s(P) + s(P') = S(PP')$.

The post-election bargaining game has (potentially) an infinite number of periods, with discounting of future payoffs using a common discount factor δ. As is standard, we will consider the limit as $\delta \to 1$. In period $t = 1, 2, 3, \ldots$, party P is chosen to make a proposal with probability $\phi_P(t)$. The function ϕ_P is called the *recognition rule* or *protocol*. The proposal is made to another party P', who responds by accepting or rejecting. If P' accepts then the game ends and the proposal is implemented. If P' rejects then the bargaining game moves to the next period. The infinite horizon specification is natural, in the sense that the parties will behave as if there is no pre-set deadline for post-election bargaining. The party who is recognized to make the very first proposal, at $t = 1$, is called the *formateur*. As a tie-breaking rule, a responder accepts an offer if he is indifferent. This assumption, which is standard in the bargaining literature, guarantees that the proposer has a best response. Otherwise, the proposer could break the indifference by offering ε more, with ε arbitrarily small. (The only exception is when the proposer is already offering *all* of the surplus — in this case, the responder can randomize and the indifference cannot be broken by offering more.)

The protocol will in general depend on the outcome of the election: a larger party is more likely to be recognized. We consider two different protocols. In the first protocol, the biggest party (i.e., P such that $n(P) > n(P')$ for all $P' \neq P$) makes the first proposal, followed by the second biggest, etc. (Tie-breaking when $n(P) = n(P')$ is discussed later, but plays no substantive role.) Formally, $\phi_P(t) = 1$ if either $t = 1, 4, 7, \ldots$ and P is the party with the largest seat share, or $t = 2, 5, 8, \ldots$ and P is the party with the second largest seat share, or $t = 3, 6, 9, \ldots$ and P is the smallest party in terms of seat share. We call this the *Austen-Smith and Banks (ASB) protocol*. In the second protocol, the probability of being recognized in each period is directly proportional to the seat shares in parliament: $\phi_P(t) = n(P)$ for all t. We call this the *random recognition protocol* (Baron and Ferejohn, 1989; Diermeier and Merlo 2004).[17]

Equilibrium Post-Election Coalition Formation

In this section, we characterize the stationary subgame perfect equilibrium (SSPE) outcomes for the ASB and the random recognition protocols.

ASB Protocol

In the ASB protocol, the outcome of the elections fully determines the order of proposers. If $n(P') > n(P'') > n(P''')$, then party P' makes the first proposal. If the proposal is rejected, P'' makes a proposal. If this is rejected, P''' makes a proposal. If this is rejected, we go to the next round, where again P' starts by making a proposal. Play continues until a proposal is accepted. Each proposal takes one period, and a discount factor δ applies to each period. Periods 1–3 constitute round 1, periods 4–6 constitute round 2, etc. Each round uses the same fixed order P', P'', P'''. With a slight abuse of terminology, we call this *ordering* the bargaining protocol. In SSPE, defined for this protocol, stationarity means behavior is the same in each round, independently of what happened in previous rounds.[18]

Let $\lambda \in \{1, 2\}$ denote the number of periods which L has to wait to make an offer after rejecting an offer from M. If M's proposal is rejected, then the next proposal is made by L if $\lambda = 1$, but by R if $\lambda = 2$. Notice that λ is determined by the election results, e.g., if $n(M) > n(L) > n(R)$ then the bargaining protocol is MLR so $\lambda = 1$. The bargaining strength of L vis-a-vis

---

In a previous version we also included a *sequential offers protocol*, in which the rejector in period t made a proposal in period $t + 1$. It did not generate any new insight so we do not consider it here.

[18] Austen-Smith and Banks (1988) originally assumed each party would only get *one* chance to make a proposal; in our terminology, there was only one round. They found that the largest and smallest party would always form the government. We allow potentially an infinite number of rounds and make a somewhat different prediction.

Pre-electoral Coalitions and Post-election Bargaining 15

M is lower, the longer L has to wait to make an offer after rejecting an offer from M. Thus, L is strong vis-a-vis M if $\lambda = 1$, and this is the only case in which R has any hope of joining a coalition government. The following result is proved in the Appendix.

Proposition 1 *For δ close to 1, the ASB bargaining game has a unique SSPE outcome. The MR coalition forms if $S(MR) > S(LM)/3$ and the bargaining protocol is either MLR or RML. Otherwise, the LM coalition forms. Whichever coalition forms, as $\delta \to 1$, player M's share of the rents from office converges to*

$$s^\lambda(M) \equiv \max\left\{\frac{\lambda}{3}S(LM),\ S(MR)\right\} \tag{1}$$

(where λ denotes the number of periods L has to wait to make an offer, if M's offer is rejected).

We sketch the intuition for the result. Notice that if the bargaining protocol is either MLR or RML, then L is not the formateur, and $\lambda = 1$.

Since the LR coalition is ruled out, in equilibrium M will either form a coalition with L or with R. Since $S(LM) > S(MR)$, whenever L proposes, L will surely make an offer to M which is sufficient to get acceptance. According to the ASB protocol, the largest party is the formateur (i.e., makes the very first proposal). Thus, if L is the largest party then the LM coalition forms in equilibrium. However, if the formateur is either M or R, the analysis is more subtle.

Suppose the protocol is MRL, so M is formateur and $\lambda = 2$. We claim the LM coalition forms in equilibrium. To see this, suppose — to derive a contradiction — the MR coalition forms. Player M can get at most $S(MR)$ from this. In period 3 player L will offer M at most $\delta S(MR)$ since (by stationarity) M gets at most $S(MR)$ next round.[19] Therefore, in period 2 player R can induce M to accept by offering $\delta^2 S(MR)$, and L is left out with a negative payoff $-x_L(MR)$. Thus, sequential rationality forces L to accept any offer from M in period 1. Since $S(LM) > S(MR)$, player M will certainly propose to L, which contradicts the hypothesis that MR forms. If the protocol is RLM, the same argument goes through. The argument

---

[19] Player M would *strictly* prefer to accept $\delta S(MR) + \varepsilon$, for any $\varepsilon > 0$, no matter how small. We assume, to guarantee that best responses exist, that he accepts also when indifferent.

uses $\lambda = 2$, so that if L rejects M's offer, player R intervenes with an offer before it is L 's turn.

Now suppose the protocol is MLR. Again, M is formateur but now $\lambda = 1$. Now L can more easily reject a proposal from M because L can immediately counter-offer. Thus, L's bargaining position vis-a-vis M is strong. But R's position is weak; he must accept any offer from M because if he rejects then L will make the next proposal and R is left out. In view of this, M prefers to propose to R if $S(MR)$ is not too small. Intuitively, although LM generates the biggest surplus, R is willing to give M a larger slice of a smaller cake. If the protocol is RML, the same argument goes through. (Of course, if L is the formateur and so can preempt all other proposals, or if $S(MR)$ is small enough to make R irrelevant, then LM must form even if $\lambda = 1$.)

Random Recognition Protocol

We now characterize the SSPE for the random recognition protocol. In each period, recognition probabilities are given by the seat shares $n(L)$, $n(M)$ and $n(R)$. Stationarity means behavior in each period is independent of what happened in past periods (history independence in the standard sense). We are primarily interested in equilibrium outcomes when δ is close to 1. The SSPE is, in general, not in pure strategies. The mixing is between acceptance and rejection (unlike, for example, Ray 2008). However, as $\delta \to 1$, the mixing becomes degenerate and the two closest parties, L and M, form a government. The following result is proved the Appendix.

Proposition 2 *For δ close to 1, the bargaining game with random recognition has a unique SSPE outcome. As $\delta \to 1$ the LM coalition always forms and player M's share of the rents from office converges to*

$$\max \left\{ \frac{n(M)}{n(L) + n(M)} S(LM), \ S(MR) \right\}.$$

We sketch the intuition for the result, retaining the notation $\phi_P \equiv n(P) > 0$ for the recognition probabilities. Fix an SSPE, and let s_P denote the equilibrium expected payoff of player $P \in \{L, M, R\}$.[20] By stationarity, s_P is the continuation payoff P expects in period $t + 1$ if the period t offer

---

[20] Party P's utility from seat shares, $\alpha n(P)$, is unaffected by the ex post bargaining as this utility is not transferable ex post. We do not include this constant in the continuation payoff for convenience.

is rejected; this does not depend on t, on who made the offer or rejected the offer, or any other aspects of past behavior. Notice that if player P is invited to join a coalition government and offered a share δs_P of the rents from office, then he is indifferent between accepting or rejecting (because if he rejects he expects to get the payoff s_P next period).

Proposals specify *non-negative* shares of the surplus generated by the government (the rents from office). There are no side-payments outside this. Therefore, the smallest amount a player can get in a coalition government is 0. The most he can get is all of the surplus, so $s_P \leq S(LM)$ for $P \in \{L, M\}$ and $s_R \leq S(MR)$. In the Appendix, we show $s_P > 0$ for each $P \in \{L, M\}$ in any SSPE.

Suppose in equilibrium R cannot make an acceptable offer to M. Now, essentially, the bargaining is between L and M, with no agreement reached in periods where R is recognized. Bargaining power in this bilateral bargaining is directly related to the recognition probabilities. Thus,

$$s_M \to \frac{\phi_M}{\phi_M + \phi_L} S(LM)$$

as $\delta \to 1$. If

$$S(MR) < \frac{\phi_M}{\phi_M + \phi_L} S(LM)$$

then this indeed gives us an equilibrium. However, if the inequality is reversed, then there is enough surplus in the MR coalition that R could intervene with an acceptable offer to M, contradicting our hypothesis.

Suppose it is in fact the case that

$$S(MR) > \frac{\phi_M}{\phi_M + \phi_L} S(LM).$$

Now we know that R must be able to make an acceptable offer to M. In a pure strategy equilibrium, player M accepts with probability 1. Player M is the only player who participates in every agreement and as $\delta \to 1$, his loss from not being recognized becomes lower and lower, as does L's payoff when recognized. Now the negative payoff L will suffer if R is recognized makes L willing to accept 0 in a coalition with M, but this implies $s_L \leq 0$. This means a pure strategy SSPE does not exist in this case, since we know that $s_L > 0$ must hold. Therefore, we must allow randomization in equilibrium.[21]

---

[21] The intuition behind non-existence of pure-strategy SSPE can be seen in a simpler setting without policy preferences, where any coalition government would generate the same surplus

In general, randomization can either be in choice of partners as a proposer, or in deciding to accept or reject as a responder. Consider the first possibility. It follows directly from our previous discussion that this is impossible; M is the only player who can randomize (since the other players can choose only M) and any randomization by M as proposer will drive L's expected payoff even lower (in our earlier discussion, M was proposing to L with probability 1). The only possible stationary equilibrium must have M randomizing between accepting and rejecting offers. Clearly, this cannot apply to offers from L, because $\delta s_M < S(LM)$ and L can force M to accept with probability 1 by offering slightly more than δs_M. Therefore, M must instead randomize between accepting or rejecting R's offer. This pins down the offer by R; it must be that R offers M the whole surplus $S(MR)$ (so R cannot induce M to accept with probability 1 by slightly raising the offer). It turns out (to maintain $s_L > 0$) that the offer is in fact accepted with a probability that goes to 0 as $\delta \to 1$, so R essentially never participates in government, although his presence at the bargaining table influences how L and M split the surplus.

Ex Ante Bargaining

A coalition formed before the election is called an ex ante coalition. Propositions 1 and 2 establish the post-election outcome in the absence of any ex ante coalitions. (We assume δ is close enough to 1 to make it legitimate to consider the limit as $\delta \to 1$.) An ex ante agreement influences the post-election negotiations by changing the number of seats the parties get in the election. Seats are translated into payoffs in a non-linear way, via the post-election recognition rule, and there is a discontinuity at seat share $1/2$ (since the majority party forms a one-party government and gets all the rents from office). This non-transferability of utility, and discontinuity at $1/2$, implies an ex ante coalition-formation problem which is somewhat non-standard. An ex ante two-party coalition is said to be *viable* if there exists

---

and there would be no externalities. The only heterogeneity would come from the ϕ_P. Suppose in this case the equilibrium payoffs are ordered in the same way as ϕ_P, and suppose this order is LMR. If M has a strictly higher continuation payoff than R, then both L and M, as proposers, will choose R, who will be in every coalition and will therefore have very high payoff, contradicting the supposed equilibrium configuration. To avoid the contradiction, there must be randomization in equilibrium, to ensure that at least two of the players have the same equilibrium payoff.

a feasible ex ante agreement which makes *both* coalition partners *strictly* better off, compared to the outcome with no ex ante coalitions. Obviously, any viable ex ante coalition must include party M.

Ex ante bargaining has a well-defined deadline, as negotiations must be concluded before the election. This introduces a kind of friction to bargaining. A deterministic protocol would be biased in favor of the party that makes the *last* proposal before the deadline, because this would be a take-it-or-leave-it offer. Following Battaglini and Coate (2007, 2008) among others, we consider instead finite-horizon bargaining with a random-recognition protocol.

The ex ante bargaining game consists of T periods, where $1 \leq T < \infty$. In each period one party is recognized to make a proposal. Specifically, party P is recognized with probability $\pi_P > 0$, where $\pi_L + \pi_M + \pi_R = 1$. The proposer P suggests an ex ante agreement to another party P', which P' accepts or rejects. If P' accepts, the ex ante agreement is in force during the election campaign. If it is rejected, we proceed to the next period. But if no agreement is reached in the final period, period T, then the ex ante bargaining game is terminated and no ex ante coalition is formed.

Since we know from previous sections what the outcome will be if there is no ex ante coalition, the ex ante bargaining game can be solved by backward induction from the last period, period T. If party $P \in \{L, R\}$ is recognized to make a (take-it-or-leave-it) proposal in period T, then P will propose a coalition with M if it is viable, in which case M will accept (but if the coalition with M is not viable, no ex ante coalition forms). If instead M is recognized in period T, then: if there is only one viable coalition he will propose it (and it will be accepted); if both the LM and MR coalitions are viable then he will propose the one that makes him better off (and it will be accepted); if no coalition is viable then no ex ante coalition forms.[22]

Thus, if there is only one period, $T = 1$, then an ex ante coalition forms with positive probability in equilibrium if and only if there exists a viable ex ante coalition. Moreover, any viable coalition has positive probability of

---

[22] To simplify the presentation, we are assuming a particular tie-breaking rule. A *null* ex ante agreement would be an agreement not to transfer any seats. To rule out this triviality, we assume that if the parties cannot strictly improve their positions, then they form *no* ex ante coalition, rather than making a null agreement. If party P strictly prefers some agreement, and party P' is indifferent, the indifference can typically be broken by giving ε more to P'. However, in a non-generic knife-edge case, offering ε more may be impossible because the only way to make transfers is via seat-shares. For simplicity, we assume this knife-edge case does not occur. Alternatively, we could assume some tie-breaking rule for this case as well.

forming. In the next two sections we investigate which ex ante coalitions are viable under different voting systems, thus obtaining a complete characterization of equilibrium ex ante coalitions for $T = 1$. The case where T is large is considered in the section entitled "Ex Ante Bargaining with Many Periods".

Viable Ex Ante Coalitions with Proportional Representation

In this section, we consider proportional representation (PR). If there is no ex ante agreement, then party P's vote share is $n(P) = v(P)$. An ex ante coalition PP' is a joint national list for parties P and P'. We assume all P supporters and all P' supporters vote for the joint PP' list, hence the list wins $v(P) + v(P')$ seats. The ex ante agreement allocates these seats among the two parties, by specifying how many (and in which order) candidates from each party appear on the joint list. (If candidates from P and P' alternate on the joint list, then each party gets half of the $v(P) + v(P')$ seats, but unequal divisions are attainable by putting more candidates from one party on the list, or putting them higher up.) Again, there is nothing else on the table ex ante.

ASB Bargaining ex post

There are two cases to consider.

Case 1: In the absence of any ex ante agreement, L and M would form a coalition government ex post.

Proposition 1 gives the conditions under which Case 1 occurs. In this case, it is impossible that L and M have a viable ex ante coalition. Indeed, L and M cannot increase their total number of seats by a joint list under PR, and (by definition of Case 1) they would form a government even with no ex ante agreement, so both parties cannot be strictly better off with a joint list. The only possible viable ex ante coalition is a joint list involving M and R, which would win $v(M) + v(R)$ seats.

There are three ways the joint MR list could be viable: it could give M its own majority, or it could enable an MR government to form ex post, or it could simply increase M's bargaining power vis-a-vis L by changing λ. The following proposition summarizes the conditions under which the MR joint list is viable for any of these reasons (the proof is given in the Appendix). Let s^λ be defined as in Equation (1), and let Δ denote the minimum number of seats that M needs to transfer to R to change λ from 1 to 2. (For example, if $v(L) > v(R) > v(M)$ then $\Delta = v(L) - v(R)$.)

Proposition 3 *Assume PR and ASB bargaining ex post. In Case 1, LM is not a viable ex ante coalition, but MR is viable if and only if at least one of the following five conditions is satisfied:*
(i)

$$x_R(LM) - x_R(M) > \alpha(0.5 - v(M)).$$

(ii) $v(M) > v(R) > v(L)$ *and*

$$S(MR) - s^1(M) + x_R(LM) > \alpha(v(R) - v(L)) > s^2(M) - s^1(M).$$

(iii) $v(L) > v(M) > v(R)$ *and*

$$S(MR) - s^1(M) + x_R(LM) > \alpha(v(L) - v(M)) > s^2(M) - s^1(M).$$

(iv) $v(L) > v(R) > v(M)$ *and*

$$S(MR) - s^1(M) + x_R(LM) > \alpha(v(R) - v(L)).$$

(v) $\lambda = 1$ *and*
$$s^2(M) - s^1(M) > \alpha\Delta.$$

A corollary to Proposition 3 is that if the ex ante bargaining game has $T = 1$, then the LM coalition never forms ex ante (since only viable coalitions can form); the MR coalition forms ex ante with positive probability if and only if at least one of the five conditions listed in the proposition holds. Depending on which of the conditions holds, the MR ex ante coalition can be followed by either a one-party government, or an LM or MR coalition government (for details see the Appendix).

Case 2: In the absence of any ex ante agreement, M and R would form a coalition government ex post.

Proposition 1 gives the conditions under which Case 2 occurs: $S(MR) > S(LM)/3$ and the bargaining protocol (in the absence of ex ante agreements) is either MLR or RML. Here, M and R cannot have a viable ex ante coalition (for the same reason that LM could not be viable in Case 1). But a joint list involving L and M might be viable, for two possible reasons: it could give M its own majority, or it could enable the LM government to form ex post (either by transferring seats from M to L or from L to M).

The following proposition summarizes the conditions under which the LM joint list is viable (the proof is given in the Appendix). Let Δ' denote the minimum number of seats that L must transfer to M in order to change the coalition government from MR to LM, and let Δ'' denote the minimum number of seats that M must transfer to L in order to achieve the same outcome.

Proposition 4 *Assume PR and ASB bargaining ex post. In Case 2, MR is not a viable ex ante coalition, but LM is viable if and only if at least one of the following three conditions is satisfied:*
(i)
$$x_L(MR) - x_L(M) > \alpha(0.5 - v(M)).$$
(ii)
$$S(LM) - s^2(M) + x_L(MR) > \alpha\Delta'.$$
(iii)
$$s^2(M) - s^1(M) > \alpha\Delta''.$$

As before, we have a corollary: if the ex ante bargaining game has $T = 1$ then the MR coalition never forms ex ante; the LM coalition forms ex ante with positive probability if and only if at least one of the conditions listed in the proposition holds. If Condition (i) holds, then the LM ex ante coalition is followed by a one-party government ex post. If Condition (i) does not hold but either (ii) or (iii) holds, then the LM ex ante partners will go on to form a coalition government ex post (for details, see the Appendix).

Random Recognition ex post

With the ASB protocol, as discussed in the previous section, player M can *give up* seats in an ex ante agreement in a way which changes the ex post protocol *in his own favor* (changes λ from 1 to 2). In contrast, with random recognition the only way M can increase his ex post bargaining power is by increasing his own seat share $n(M)$ (see Proposition 2). But the other parties would never agree to transfer seats to M just to increase M's ex post bargaining power. (There are no side payments ex ante which M can use to buy seats; seats are the only currency.) Thus, under random recognition an ex ante coalition cannot form simply to manipulate the bargaining power within a given government. It could, however, produce a majority government. Proposition 2 implies that if no party has its own majority then L and M form a coalition government, and R's payoff is $\alpha v(R) - x_R(LM)$. As before, in this situation the LM coalition is not viable ex ante. If, however, M and R form an ex ante coalition wherein R gives up $0.5 - v(M)$ seats to M, then M gets its own majority and R's payoff is

$$\alpha[v(R) - (0.5 - v(M))] - x_R(M).$$

Clearly, this agreement benefits M. It benefits R if the gain from the reduced externality, $x_R(LM) - x_R(M)$, exceeds the value of the lost seats, $\alpha(0.5 - v(M))$. Thus, an MR ex ante coalition might form in order to reduce the externality on R and give M a majority. We get the following result.

Proposition 5 *Assume PR and random recognition bargaining ex post. LM is not a viable ex ante coalition, but MR is viable if and only if*

$$x_R(LM) - x_R(M) > \alpha(0.5 - v(M)).$$

The MR ex ante coalition, if it forms, always leads to an M majority government.

A corollary to this result is that if the ex ante bargaining game has $T = 1$, then the MR coalition has positive probability of forming ex ante if and only if $x_R(LM) - x_R(M) > \alpha(0.5 - v(M))$. The MR ex ante coalition enables M to form a majority government ex post. The LM coalition never forms ex ante. With random recognition, there is only one motive for ex ante coalition formation, namely, the externality motive, so there is less scope for ex ante agreements under the random recognition rule than under the ASB protocol.

Viable Ex Ante Coalitions with Plurality Voting

In the case of plurality voting, an ex ante agreement divides up the districts, and specifies in which districts each party should run and where it should drop out of the race. This is the *only* item on the table for ex ante negotiations. In particular, an ex ante coalition cannot make commitments about behavior in the post-election stage.

With no ex ante agreement, party P's vote share would be $w(P)$. If a party drops out of a district, its supporters will vote for the ideologically closest party. Thus, if M drops out of a district, the M-supporters in the district vote for L; if L drops out the L-supporters vote for M. In either case, R only wins the district if his supporters form a majority. Because of this, if L and M form an ex ante coalition and divide up all the districts, they will reduce R's seat share from $w(R)$ to $z(R)$. The remaining $w(R) - z(R) > 0$ seats can be redistributed within the LM coalition (by dividing up the districts appropriately) to make both L and M better off. In other words, L and M can always benefit by not splitting the vote. This is true whether ex post bargaining is random recognition or ASB. Thus, we get:

Proposition 6 *Under PV, the LM ex ante coalition is always viable.*

Thus, if the ex ante bargaining game has $T = 1$, then the LM coalition has positive probability of forming ex ante. It remains to consider if the MR coalition can be viable.

With plurality voting, there is an asymmetry in the way parties can transfer seats. R can transfer seats to M, because if R drops out of a district, the R-supporters vote for M, and M wins the district unless the L-supporters form a majority. But M cannot transfer any seats to R, because if M drops out of a district, the M-supporters vote for L (since, ideologically, M is closer to L).[23] Despite this non-transferability, it is possible for the MR coalition to be viable. To study this, we need to be more specific about the ex post bargaining protocol.

ASB Bargaining ex post

Apart from not splitting the vote, there are two other possible motives for an ex ante coalition: to influence which government will form, and to manipulate the bargaining power within the government. It is again useful to distinguish two cases, depending on what would happen ex post in the absence of ex ante agreements. The precise conditions for each case are derived in the Appendix. Here we provide a brief discussion of the intuition and state the main results.

Case 1: In the absence of ex ante agreement, L and M would form a coalition government. (The conditions for this to happen are given in Proposition 1.)

---

[23] Under PR, M can transfer seats to R by putting R-candidates higher on the list.

We distinguish two sub-cases.

Sub-case 1a: Party R is ex post irrelevant in the sense that $S(MR) < S(LM)/3$.

For the MR coalition to be viable ex ante, player R must be made better off than if there is no ex ante coalition. By Proposition 1, in Case 1a, R will never be part of a coalition government. Moreover, recall that M cannot transfer seats to R under PV. The only way the ex ante agreement can make R better off is if it leads to a majority government formed by M (rather than an LM coalition government). For this to happen, R must drop out of some districts in order to raise M's seat share to $1/2$. For this to be worthwhile for R, the negative externality the LM coalition government imposes on R must be sufficiently big. The precise condition is given in the following proposition, proved in the Appendix:

Proposition 7 *Assume PV and ASB bargaining ex post. In Case 1a, the MR ex ante coalition is viable if and only if*

$$x_R(LM) - x_R(M) > \alpha w(R)\frac{0.5 - w(M)}{1 - z(L) - w(M)}. \tag{2}$$

A corollary is that if the ex ante bargaining game has $T = 1$, then the MR coalition has positive probability of forming ex ante if and only if the inequality stated in Proposition 7 holds. The MR ex ante coalition, if it forms, is always followed by a one party government.

Sub-case 1b: Party R is ex post relevant in the sense that $S(MR) > S(LM)/3$.

In Case 1b, there are more ways for the MR ex ante coalition to be viable. In particular, it might lead to the MR coalition government. Because of this, the conditions for the MR coalition to be viable are more involved than for Case 1a. Details are relegated to the Appendix. (Incomplete seat sharing arrangements are possible.)

Finally, the simplest case.

Case 2: In the absence of ex ante agreement, M and R would form a coalition government ex post. (Proposition 1 gives the conditions for this to happen.)

By a familiar argument, in Case 2 the MR coalition cannot be viable ex ante (and hence never forms if $T = 1$).

Random Recognition ex post

According to Proposition 2, with random recognition party R is always ex post irrelevant, in the sense that R cannot possibly be part of any coalition government. The analysis is, therefore, similar to Case 1a, i.e., when R is irrelevant. The MR ex ante coalition is viable only if the negative externality the LM coalition government imposes on R is sufficiently big, which is true if and only if the inequality stated in Proposition 7 (Inequality 2) holds. We thus have the following result:

Proposition 8 *Assume PV and random recognition ex post. The MR ex ante coalition is viable if and only if Inequality 2 holds.*

Hence, if $T = 1$, the MR coalition has positive probability of forming ex ante if and only if Inequality 2 holds, and this ex ante coalition is always followed by a single party government.

With random recognition, the only motives for ex ante coalition formation are the externality motive and not splitting the vote. In contrast, as we have seen, with the ASB protocol an ex ante coalition could also form in order to manipulate ex post bargaining power. Therefore, with PV as well as with PR, there is less scope for ex ante agreements under the random recognition rule than under the ASB protocol.

Ex Ante Bargaining with Many Periods

In our model, M is the essential player who must be part of any coalition (as the LR coalition never forms). However, bargaining frictions may prevent M from taking full advantage of his favorable position. Ex post, L and R have bargaining power due to their ability to delay government formation, which lowers M's payoff via discounting. Ex ante, there is a different kind of friction caused by the finite T.

The sections "Viable Ex Ante Coalitions with Proportional Representation" and "Viable Ex Ante Coalitions with Plurality Voting" characterized the set of equilibrium ex ante coalitions for $T = 1$. Consider two cases. First, both LM and MR might be viable ex ante coalitions (in the sense of the "Ex ante Bargaining" section). Since with $T = 1$ the formateur makes a take-it-or-leave-it offer, M will accept any proposal which is better than no ex ante agreement. Thus, if both LM and MR are viable, player M is not guaranteed to get the ex ante coalition he prefers. If L is recognized, L proposes LM which is accepted; if R is recognized, R proposes MR which is accepted. Realized payoffs depend on the luck of he draw, i.e., on who is formateur, and ex ante bargaining power is proportional to the probability of being recognized. The second case is where only one coalition, say MR, is viable ex ante. In this case, with $T = 1$, coalition MR will form ex ante if either M or R is recognized, but if L is recognized no ex ante coalition forms.

Now we will consider what happens when T is large. As T increases, there are more and more opportunities for offers and counter-offers, and M can better exploit his favorable bargaining position. We will show that M's equilibrium payoff always increases monotonically in T, and study how this changes the equilibrium outcome in the two cases discussed in the previous paragraph (either both LM and MR are viable coalitions, or just one of them).

The ex ante bargaining has a natural deadline: the time papers must be filed for the election. The interpretation of a large T is the following. Ex ante bargaining takes place during a week, say, before the deadline. The week is divided into T periods when offers can be made. A large T means frictions to bargaining vanish, as potentially a very large number of proposals can be made during the week (as the interval between successive offers is very brief). Reaching an agreement early in the week is no more advantageous than reaching an agreement later in the week, as payoffs only accrue after the election. Thus, we assume no discounting between periods. (For any finite T, the equilibrium with no discounting would be the limit of discounted equilibria when the discount factor goes to 1.)

From the previous sections, we know how the players will behave in the last period, period T. Equilibrium behavior in preceding periods is obtained by backward induction. Let $u_j(M)$ denote M's expected payoff at the very beginning of period $T - j$, assuming the players have not reached any agreement before this period. That is, $u_j(M)$ is player M's expected payoff with j periods left to bargain before the deadline. Let $u_j(M|P)$ denote M's expected payoff if $P \in \{L, M, R\}$ is recognized to make a proposal with j periods left. Since party P is recognized with probability π_P, we have

$$u_j(M) = \pi_L u_j(M|L) + \pi_M u_j(M|M) + \pi_R u_j(M|R). \tag{3}$$

Now consider the preceding period, period $T - (j + 1)$. Since any coalition must involve M, player M can always make sure there is no agreement in period $T - (j + 1)$, in which case we move to period $T - j$, where player M's expected payoff will be $u_j(M)$. Thus, in period $T - (j + 1)$, player M will never accept anything less than $u_j(M)$. That is, regardless of which party is recognized in period $T - (j + 1)$, player M's expected payoff must be at least $u_j(M)$:

$$u_{j+1}(M|P) \geq u_j(M). \tag{4}$$

Therefore,

$$u_{j+1}(M) = \pi_L u_{j+1}(M|L) + \pi_M u_{j+1}(M|M) + \pi_R u_{j+1}(M|R) \geq u_j(M).$$

Thus, $u_j(M)$ is increasing in j. The more periods left to bargain, the better off is M, as he can better exploit his position as the player who must participate in any agreement. Notice that player M's equilibrium payoff equals his expected payoff at the very beginning of the ex ante bargaining, $u_T(M)$. Since $u_T(M)$ is increasing in T, player M is better off the more opportunities there are for offers and counter offers, i.e., the larger is T.

Since payoffs are bounded above by the size of the available surplus, the increasing sequence $\{u_j(M)\}_{j=1}^{\infty}$ must approach a limit $u^*(M)$ as j becomes large. That is,

$$\lim_{j \to \infty} u_j(M) = u^*(M).$$

But now Conditions (3) and (4) imply that $u_j(M|P)$ also converges to $u^*(M)$ as j become large. Thus, we have proved:

Proposition 9 *There is $u^*(M)$ such that for any $P \in \{L, M, R\}$, $u_j(M|P) \to u^*(M)$ as $j \to \infty$.*

Ex ante coalitions have three motives for forming: (i) to avoid splitting the vote in PV; (ii) to change the order of proposers when ASB is used ex post; and (iii) to give M its own majority if α is small and externalities are relatively large. If T is small, and more than one motive is present, the luck of the draw determines which motive prevails. But when T is large, Proposition 9 implies the indeterminacy disappears and M will be able to exploit his position as the essential player, by playing L and R off against each other. More precisely, M's payoff depends neither on who happens to be recognized, nor on T itself.

We now consider the implications of Proposition 9 for the case where both LM and MR are viable coalitions in the sense of the "Ex Ante Bargaining" section. For simplicity, consider the specific case of plurality voting with random recognition ex post. By Proposition 2, if there is no ex ante agreement, L and M form the government ex post. Suppose Inequality 2 holds, so by Propositions 6 and 8, both LM and MR are viable coalitions: LM due to motive (i) and MR due to motive (iii). If $T = 1$, then if R were recognized the MR coalition would form ex ante, and M would form a one-party government ex post (which benefits R, as it prevents the LM government). If M were recognized, he would of course do no worse, i.e., he would get at least 50% of the seats in parliament and all of the rents from office, so his payoff would be at least $0.5\alpha + S$. However, if instead L were recognized to make the take-it-or-leave-it offer, then LM would form both ex ante and ex post, which would give a much lower payoff for M.

We now show that when T is large, M must form a one-party government ex post. It suffices to show that $u^*(M) \geq 0.5\alpha + S$, where $u^*(M)$ is defined as in Proposition 9. Since $u_j(M|P) \to u^*(M)$, for any P, the probability of M forming a one-party government must be close to 1 if $u^*(M) \geq 0.5\alpha + S$.[24]

To obtain a contradiction, suppose $u^*(M) < 0.5\alpha + S$. Suppose there are j periods left. For sufficiently large j, M cannot form a majority government,[25] and the probability of the LM government forming must be close to 1. But Inequality 2 is exactly the condition for R to prefer to give up sufficiently many seats to give M its own majority, rather than suffering the externality from LM. Therefore, if M proposes such a coalition, R must accept. But then there is a proposal M can make, which gives him at least $0.5\alpha + S$, so $u_j(M|M) \geq 0.5\alpha + S > u^*(M)$ in contradiction of Proposition 9. This contradiction proves that $u^*(M) \geq 0.5\alpha + S$. Thus, when T is large, M surely forms a one-party government ex post.

Since $S(LM) > S(MR)$, L and M are natural coalition partners. Within the LM coalition government, M's bargaining position is strong if his ideological differences with R are not too big, i.e., if $S(MR)$ is not too small. Indeed, Propositions 1 and 2 show how M benefits from an increase in $S(MR)$, even if the LM government forms. With ASB bargaining, the MR

---

[24] With less than 50% of the seats, player M would share $S(LM) < S$ with L. Therefore, M's payoff would be much less than $0.5\alpha + S$.

[25] Indeed, if P's proposal were such that M gets a majority in parliament, then M's payoff would be at least $0.5\alpha + S$, so we would have $u_j(M|P) \geq 0.5\alpha + S > u^*(M)$. For large enough j, this would contradict Proposition 9.

government could even form ex post if $S(MR)$ is close to $S(LM)$. However, M benefits in a different way if L and R become more ideologically distant from each other, because externalities will be large and Inequality 2 will be more likely to hold. In this case, R is willing to cooperate with M to ensure M has its own majority. As we saw in the previous paragraph, M will then be guaranteed to form a majority government if T is large. Thus, while M's bargaining position vis-a-vis L is made weaker if ideological polarization eliminates MR as a possible coalition government, polarization can benefit M in a different way: he is more likely to form a one-party government if ideology is so important that the main concerns of L and R are to exclude each other from the government.

Finally, we briefly consider the case where only one coalition, say MR, is viable in the sense of the "Ex Ante Bargaining" section. If $T = 1$ the LM ex ante coalition cannot form since it is not viable by hypothesis. Now suppose T is large. If LM never forms ex ante, then MR must form with probability close to 1, since M and R would eventually be recognized and both prefer MR to no coalition (MR is viable by hypothesis). Therefore, if M makes an ex ante proposal to L which gives L a payoff greater than what he gets when the MR ex ante coalition forms, player L must accept it. This could well be better for M than forming an ex ante coalition with R, in which case the hypothesis that LM never forms ex ante cannot be maintained. In other words, player M is better off when T is large because the LM ex ante coalition becomes an equilibrium outcome. Intuitively, player M does better when T is large because the other parties are more inclined to join an ex ante coalition, knowing that they cannot prevent their ideological opponent from joining M if they refuse.

Strategic Voter Coordination

With a continuum of voters no single vote has any impact on the election outcome. Therefore, we have so far assumed voters always vote for the candidate who is the closest in terms of ideological position, while the parties strategically manipulate the elections by forming ex ante coalitions. However, if large numbers of voters can coordinate their behavior, then they can impact the election. An interesting question is whether this could replicate the outcome of ex ante coalition formation among the parties. Morelli (2004) asked a similar question, and Indridason (2009) considered how the expected ex post coalitions in PR affect strategic voter behavior.

Suppose the ex ante coalition formation stage is replaced by a *recommendation stage*. In this stage, the three parties simultaneously tell their own supporters how to vote in the upcoming election. All P-supporters have the same payoff function as party P, so they can be assumed to follow P's recommendations.[26] With PR, the recommendation of party P allocates its $v(P)$ supporters (and hence $v(P)$ seats) among the three parties. In the case of PV, a recommendation advises supporters in each district whom they should vote for. After the recommendation stage, the election is held, followed by post-election bargaining as before. We call this three-stage game the *strategic vote coordination game*.[27]

In the strategic vote coordination game, the parties can transfer seats in ways that ex ante coalitions could not achieve in the absence of voter coordination. For instance, with PV and no coordination, we saw above that M cannot help R gain seats by dropping out of districts, because the M-supporters would vote for L rather than R. In contrast, in the strategic vote coordination game, M can recommend the M-supporters in some districts to vote for R, while R recommends that in the remaining districts the R-supporters vote for M. As a result, L would win only $z(L)$ seats, and both R and M could raise their seat shares. However, such an implicit agreement, involving a two-way transfer of votes between M and R, could in fact not be part of an equilibrium of the vote coordination game. The problem is that the strategic vote coordination game lacks even the minimal commitment of pre-electoral coalitions. Both M and R would prefer to deviate at the recommendation stage, by recommending that their voters should, in fact, vote for their own party after all.

With ex ante coalitions under PV we showed that M and L can avoid splitting the vote (and drive R's seat share down to $z(R)$) by agreeing to a two-way transfer of votes, in effect making a binding commitment not to compete in certain districts by not filing nomination papers. However, in the strategic vote coordination game there are no binding commitments: both

---

[26] Thus, P-supporters care about party P's seats in parliament, its rents from office, and the negative externality from ideologically distant parties in government. Intuitively, if a party is in government it can reward its base to an extent proportional to the amount of power it exerts. If the party is not in government, its voters instead incur a loss corresponding to an undesirable policy.

[27] Equivalently, one could assume that a representative voter controls the votes of each party. These representative voters make implicit agreements to divide up the districts, but when they get to the voting booth, they are free to violate the agreement, as they cannot be forced to vote for someone they don't like.

L and M are free to tell their voters to vote for their own party. Therefore, an equilibrium ex ante agreement to drive R down to $z(R)$ seats cannot be achieved. Due to lack of commitment, such vote coordination cannot replicate the ex ante coalitions of the original game.

We have so far discussed two-way transfers of votes. But since (as noted above) there is a larger feasible set of transfers under strategic vote coordination, we cannot even be sure that ex ante coalitions involving one-way transfers can be replicated in an equilibrium. Transfers that were not possible without coordination might now be best responses. Also, for example, in the case of ASB with $\lambda = 1$, M might want to give up seats to R to change λ to 2. However, it is not necessarily a best response here for L to do nothing; it might ask its voters in some districts to vote for M to bring λ back to 1. Of course, it is not a best response for M to give up vote share or seats if L is going to counteract it, so this too cannot be an equilibrium. Equilibrium might therefore require mixed strategies.

Suppose *both* ex ante agreements among the parties and strategic recommendations were permitted. Then the commitment issue would be resolved, if the recommendations followed the filing of nomination papers or announcement of a list, so that withdrawals from districts were enforceable. The main way this would differ from the original model without coordination is that transferring districts from M to R would become feasible even with PV, so the MR ex ante coalition could become viable. The MR ex ante coalition could even be the best option for M, and thus an equilibrium outcome (both with $T = 1$ and T large). Thus there would be potential differences in the ex ante coalition structure with and without coordination, but only under PV. Even the LR coalition, where R gains seats and L gets increased ex post bargaining power, could not be ruled out. The possibilities are too many to make it analytically tractable to characterize the equilibria.

Conclusion

Why are pre-electoral coalitions so prevalent? In reality, they frequently do not seem to represent a binding commitment to stay together after the election, much less a commitment to share the rents from office in a particular way. Still they do imply a limited kind of commitment. With national elections, it is the commitment to allocate seat shares by a joint national list. With plurality district voting, not filing papers by the deadline is a commitment not to contest a district. Our model shows how this limited form of commitment can influence coalition formation in parliamentary democracies. It illustrates the rich strategic interplay between legislative bargaining and seat-sharing arrangements in the context of legislative elections.

We characterized the (stationary) equilibria of two (infinite horizon) ex post bargaining protocols, ASB and random recognition, in the presence of externalities and compromise costs among players with different ideologies. We analyzed how the ex post bargaining protocol influences ex ante coalition formation, even though ex ante coalitions can break up ex post. Ex ante coalitions form to influence which government will form ex post, to manipulate the bargaining power within the government, and (with PV) to avoid splitting the vote. With the PV electoral system, ex ante coalitions between L and M increase their total seat shares. On the other hand, R might be willing to make an agreement which gives M its own majority. This is more likely if the parties are ideologically polarized, so that R suffers a large negative externality if L joins a coalition government. With the ASB protocol ex post, ex ante coalitions might also form in order to manipulate the order in which players are recognized in the ex post protocol. Thus, with PR, players L and M may benefit from a joint list even though it does not increase their total seat shares. Depending on the parameters, an MR coalition might instead form to ensure a majority for M.

Ex ante bargaining has a fixed deadline, so only a finite number T of proposals can be made. Discounting does not play a major role since rejections do not delay payoffs (in contrast to post-election bargaining, where a rejection delays the government formation). In general, M benefits from being ideologically fairly close to R, so $S(MR)$ is fairly large, since this makes R a more credible coalition partner, and M can play L and R off against each other. However, M also benefits from L and R being ideologically distant from each other, so each would suffer a large externality from the other forming a coalition government with M. If these externalities are important then the equilibrium outcome may be an M-party majority government.

Our model captures some complexities of real-world parliamentary democracies. A stylized fact is that pre-electoral coalitions are common, and our model suggests a number of possible reasons for this. Reasons exist even under the purest form of PR, as in Israel. In Golder's (2006a, b) sample, 87% of Israeli elections had some kind of pre-electoral coalition. Still, while reasons for pre-electoral coalitions exist under both PR and PV, the reasons are clearly stronger under PV. Golder finds evidence that the more a system diverges from PR, the higher the incidence of pre-electoral alliances.

Our model suggests several questions for empirical work. First, how does the post-election bargaining protocol influence pre-electoral coalition formation? Debus (2009) and Golder (2006b) studied how pre-electoral coalitions affect post-election government formation, but did not consider the possibility of causality running in the opposite direction, from post-election bargaining rules to pre-electoral coalition formation. Some countries rely on deterministic protocols for choosing the formateur. For example, in Greece and Bulgaria the ASB protocol is constitutionally spelled out, and in India and the U.K. it is a strong convention. In such countries, bargaining power will be highly sensitive to changes in vote shares, which suggests pre-electoral coalitions should be relatively common.

Second, what other variables influence the pre-electoral coalition formation? For instance, in our model a possible motive for pre-electoral coalitions is to make sure the median party is strong enough to form a majority government. The willingness of a smaller party to participate in such an arrangement would depend on how it trades off the value of seats in parliament against the benefit of preventing ideologically distant parties from joining the government. If seats are valuable (α high in our model), then such arrangements will be rare. The trade-off also depends on the extent to which opposition parties can influence legislation (via legislative committees, for example) and perhaps receive public campaign funds based on their performance in past elections.

Third, what determines whether pre-electoral coalitions break apart after the election? In our model, an ex ante coalition consisting of parties that are ideologically distant (M and R) could be followed by a coalition government of parties that are ideologically closer (L and M) (see Proposition 3 and its corollary). More generally, equilibrium coalitions depend on the relative ideological distance between parties, the size of the surplus for each coalition, and the post-election bargaining protocol. This could be studied empirically. If ideology is very important and parties are highly polarized, then parties are willing to concede a lot in order to prevent ideological opponents from joining the government. If bargaining frictions are small, the median party can exploit this polarization and (in our model) even achieve its own majority. On the other hand, polarization would mean that R is also quite distant from M and this would reduce $S(MR)$. Therefore, M's bargaining power drops within the LM coalition, as R is no longer a viable coalition partner.

Conversely, if a coalition government forms, the median party is predicted to get a larger share of the rents from office (say, more cabinet posts) if polarization is low enough and it has other viable coalition partners. Testing this empirically would require some measure of ideological closeness, based, for example, on the one-dimensional spatial model.[28]

A different set of empirical issues relate more specifically to post-election bargaining and distribution of surplus. With random recognition, equilibrium surplus shares are not necessarily proportional to seat shares (contrary to Gamson's law), because M's surplus is bounded below by $S(MR)$ (Proposition 2). Intuitively, if the LM government forms, R gets no rents from office and in addition suffers a negative externality. Therefore, R is willing to join a coalition government with M even if M takes all of $S(MR)$, and so M will not accept less than $S(MR)$ in the LM government. Golder and Thomas (2011) and Indridason (2010) have in fact found evidence of systematic departures from Gamson's law, and our model suggests that the bargaining protocol as well as the relative viability of bargaining partners (the relative sizes of $S(LM)$ and $S(MR)$) could be explanatory variables.

With a deterministic ex post protocol (such as ASB), the role of R is much more involved than simply putting a lower bound on M's payoff. For instance, suppose M is the largest and R the smallest party, so the bargaining protocol is MLR. If M rejects a proposal by L, party M cannot immediately make a counter-proposal, but has to wait for R to make an intervening proposal. This reduces M's bargaining power vis-a-vis L, hence M would get a relatively small share of the surplus $S(LM)$. Party M may prefer to form a coalition government with R and get the lion's share of $S(MR)$. Therefore, the smallest party R, which is ideologically rather distant from M, could nevertheless join a coalition government (Proposition 1). Even if the LM government forms, the surplus shares of L and M are influenced by the existence of R. Thus, with ASB-type bargaining, smaller parties in theory have a lot of influence (and thus reason to exist) even under PV (contrary to Duverger's law). In India (which uses PV), there are many small parties which do seem to influence post-election outcomes.

Under PR, a key simplification of our model was that seats are proportional to vote shares with no minimum vote threshold. Extending our analysis to PR systems with minimum thresholds would introduce a new strategic issue: a small party may form a pre-electoral coalition to ensure its

---

[28] Indridason (2011) examines how polarization affects coalition formation.

representation in parliament. It would be interesting to study when larger parties would want to form such ex ante alliances. With no aggregate uncertainty, as in our model, it can be anticipated before the elections when a party will not meet the threshold. If it is L or R, this party would be willing ex ante to transfer all its votes to M, rather than the other party winning. If any two parties together have a majority, M will then form a one-party government. Now if it is M which does not meet the threshold, it will still have an advantageous pre-electoral bargaining position. Suppose R would become the majority party with no ex ante agreement. The LM coalition is viable ex ante, either leading to an LM coalition government, or L would get its own majority. Both outcomes are better for M than a majority for R. Depending on the trade-off between seats and ideology, party R might be willing to give up seats to M to prevent an LM coalition government. With T large, we expect that party M will play off L against R and easily cross the threshold for representation. Finally, it would be interesting to embed this model in a presidential system, adding a key strategic player who may have veto power.

Appendix

Stationary equilibria of the ASB bargaining game

Here, we prove Proposition 1. At the ex post stage, the parties bargain over government formation and surplus sharing. Each party's seat share $n(P)$ has been determined in the election and enters as a fixed term $\alpha n(P)$ in the payoff function. For simplicity, we omit this fixed term when deriving the payoff from the bargaining game. Notice that, because of the negative externality, a player left out of the government may get a negative payoff. But a governing party P gets a non-negative share of the rents from office and suffers no externality, and thus P (unlike the outsider) cannot get a negative payoff.

Proposition 1 is proved via three lemmas, corresponding to three different situations. Lemma A.1 deals with the case where M is the smallest party, and so is the last party to make a proposal within each round. In this case, it is quite easy to show that the LM coalition is the equilibrium outcome.

Lemmas A.2 and A.3 deal with the case where M is not the smallest party. Lemma A.2 shows that if $\lambda = 1$, then the LM coalition forms only if $S(MR) < S(LM)/3$. Lemma A.3 shows that when $\lambda = 2$, the LM coalition must form. When $\lambda = 2$, it takes some care to compute equilibrium payoffs, so the proof of Lemma A.3 is slightly longer than the others.

Throughout, we restrict attention to stationary subgame perfect equilibria (SSPE). Let s_P denote party P's equilibrium payoff. By stationarity, whenever any round is about to begin, P expects s_P. Recall that λ denotes the number of periods L has to wait to make an offer after rejecting an offer from M.

Lemma A.1 *If M is the smallest party, then the LM coalition must form. As $\delta \to 1$, player M's share of the rents from office converges to*

$$\max\left\{\frac{\lambda}{3}S(LM), S(MR)\right\}.$$

Proof: Suppose M is the smallest party, so the protocol is either LRM or RLM. We first show that the LM coalition must form. Suppose, in order to derive a contradiction, that the equilibrium outcome is the MR coalition. Since M is the smallest party, M is the last to make an offer in each round. Player L will accept any offer from M, since by rejecting, he expects to be left out in the next round (by stationarity). Accordingly, M can reject all offers and when it is his turn to propose get all of the surplus in a coalition with L, as L will accept an offer of 0. For δ close to 1, this is certainly better for M than forming a coalition with R (since $S(LM) > S(MR)$). This contradiction shows that the LM coalition must form in equilibrium.

Next, we compute M's equilibrium payoff s_M. If M makes an offer which is rejected, then a new round begins where (by stationarity) the LM coalition is expected to form, L's expected payoff is s_L, and R's expected payoff is negative (since it is left out). Therefore, R will accept any offer from M, while L will accept any offer of at least δs_L. This implies that M's expected payoff when it is his turn to propose is

$$\tilde{s}_M = \max\left\{S(LM) - \delta s_L,\ S(MR)\right\}. \tag{A.1}$$

When it is L's turn to propose, he will offer $\delta^{3-\lambda}\tilde{s}_M$ to M and M will accept.[29] This implies that L's expected payoff when it is his turn to

---

[29] This is obvious if the protocol is RLM, since then $\lambda = 2$ and M is the next to make an offer after L. If $\lambda = 1$, so the protocol is LRM, then R makes the next offer after L. But, even if R can make an acceptable offer to M, R will never offer more than $\delta\tilde{s}_M$. Therefore, L can offer $\delta^2\tilde{s}_M$.

propose is

$$\tilde{s}_L = S(LM) - \delta^{3-\lambda}\tilde{s}_M.$$

If $\lambda = 1$, so the protocol is LRM, then $s_L = \tilde{s}_L$. If $\lambda = 2$, so the protocol is RLM, then (since the game starts with R making an offer which is rejected) $s_L = \delta\tilde{s}_L$. Thus, in either case,

$$s_L = \delta^{\lambda-1}\tilde{s}_L = \delta^{\lambda-1}\left(S(LM) - \delta^{3-\lambda}\tilde{s}_M\right). \tag{A.2}$$

Substituting Equation A.2 in Equation A.1 and taking the limit as $\delta \to 1$ yields

$$\tilde{s}_M = \max\left\{\frac{1-\delta^\lambda}{1-\delta^3}S(LM),\ S(MR)\right\} \to \max\left\{\frac{\lambda}{3}S(LM),\ S(MR)\right\}.$$

Player M's equilibrium payoff s_M converges to the same limit because he is offered $\delta^{3-\lambda}\tilde{s}_M$ by L.   ∎

Lemma A.2 *Suppose the protocol is either MLR or RML. If $S(MR) < S(LM)/3$ then for δ close to 1, the LM coalition forms. If $S(MR) > S(LM)/3$ then for δ close to 1, the MR coalition forms. In either case, as $\delta \to 1$, player M's share of the rents from office converges to*

$$\max\left\{\frac{1}{3}S(LM),\ S(MR)\right\}.$$

Proof: (i) Consider the protocol MLR. If M rejects an offer from L, then M expects to get $\delta^2 s_M$. To see this, note that R is the next to propose. Either R's offer is rejected, in which case M has to wait and get s_M when the new round starts. Or else M accepts an offer from R, but this offer cannot exceed δs_M, by a standard argument (if it exceeds δs_M then R could lower the offer slightly and it would still be accepted).

When it is L's turn to make a proposal, he will offer $\delta^2 s_M$ to M and, by the usual argument, this must be accepted by M (for L could surely get acceptance by offering slightly more). Continuing the same argument, when it is M's turn to make an offer, L will accept any offer which gives him at least $\delta(S(LM) - \delta^2 s_M)$. Player R will accept even a zero offer, since a rejection means L makes an offer to M which is accepted. Therefore,

$$s_M = \max\left\{S(LM) - \delta(S(LM) - \delta^2 s_M),\ S(MR)\right\}$$

which is equivalent to

$$s_M = \max \left\{ \frac{1-\delta}{1-\delta^3} S(LM), \; S(MR) \right\} \rightarrow \max \left\{ \frac{1}{3} S(LM), \; S(MR) \right\}$$

as $\delta \rightarrow 1$. If $S(MR) > S(LM)/3$, then for δ close to one,

$$S(MR) > \frac{1-\delta}{1-\delta^3} S(LM).$$

This means M's offer goes to R. If $S(MR) < S(LM)/3$, then for δ close to one,

$$S(MR) < \frac{1-\delta}{1-\delta^3} S(LM)$$

and M's offer goes to L.

(ii) If the protocol is RML, the argument is similar and is omitted.   ∎

Lemma A.3 *Suppose the protocol is MRL or LMR. For δ close to 1, the LM coalition forms. As $\delta \rightarrow 1$, player M's share of the rents from office converges to*

$$\max \left\{ \frac{2}{3} S(LM), \; S(MR) \right\}.$$

Proof: (i) Consider the protocol MRL, where M is formateur. First, we show the LM coalition must form. Suppose — in order to obtain a contradiction — that M offers a payoff of $w \geq 0$ to R, which is accepted. This implies that if M rejects an offer from L, player M expects to get $S(MR) - w$ next period (the beginning of the new round). Therefore, L will offer him $\delta(S(MR) - w)$, and this is accepted by M. Therefore, if M rejects an offer from R, player M will get $\delta(S(MR) - w)$ next period. Hence R offers $\delta^2(S(MR) - w) < S(MR)$, and M accepts. But then L must accept any offer from M, since if he rejects he is left out. Clearly, M prefers to offer 0 to L rather than w to R, a contradiction. Thus, in each round, M makes an offer to L, not to R.

Having established that the LM coalition forms, it remains to show that M's payoff takes the familiar form.

Player M is the formateur. In equilibrium, he offers a payoff of s_L to L, which is accepted, and M himself gets $s_M = S(LM) - s_L$. If M rejects an offer from L, next period (which is the beginning of the next round) M

gets s_M. Therefore, L will offer δs_M, and M accepts. Therefore, if M rejects an offer by R, M will get δs_M next period. Hence M must accept an offer of $\delta^2 s_M + \varepsilon$ from R. If $\delta^2 s_M < S(MR)$, then such offer is feasible, and R is sure to make a proposal to M that will be accepted. In this case, L will expect to be left out if he rejects an offer from M, hence L will accept any offer, which means $s_M = S(LM)$. For δ close to 1, this contradicts $\delta^2 s_M < S(MR)$. This contradiction shows we must satisfy the condition

$$\delta^2 s_M \geq S(MR). \tag{A.3}$$

There are two possibilities.

Case α. Suppose there is strict inequality in Condition A.3. Then, R cannot make any proposal to M that M will accept. Hence, if L rejects M's offer, R cannot make any agreement with M, and L will offer δs_M two periods later, which M accepts. Thus, L will accept M's offer if and only if it exceeds $\delta^2(S(LM) - \delta s_M)$. This expression must equal $s_L = S(LM) - s_M$, which implies

$$s_M = \frac{1 - \delta^2}{1 - \delta^3} S(LM). \tag{A.4}$$

For δ close to 1, Equation (A.4) is consistent with Condition A.3 if $S(MR) < \frac{2}{3}S(LM)$, but not if $S(MR) > \frac{2}{3}S(LM)$. Thus, Case α cannot happen if $S(MR) > \frac{2}{3}S(LM)$. Suppose instead $S(MR) < \frac{2}{3}S(LM)$. Player M offers $s_L = S(LM) - s_M$ to L, and L accepts any offer of at least this much. If M's offer is rejected, in the next period M rejects all feasible offers from R. This is optimal for M, because

$$S(MR) < \delta^2 s_M = \delta^2 \frac{1 - \delta^2}{1 - \delta^3} S(LM)$$

for δ close to 1. Player L offers δs_M to M, and M accepts any offer of at least this much. These strategies form an SSPE, and $s_M \to \frac{2}{3}S(LM)$.

Case β. Suppose there is equality in Condition A.3, so

$$s_M = \frac{1}{\delta^2} S(MR). \tag{A.5}$$

In any round, M offers $s_L = S(LM) - s_M$ to L, and L accepts any offer of at least this much. If M's offer is rejected, then in the next period R offers $S(MR)$ to M. When M gets this offer, Equation (A.5) implies that he is indifferent between accepting and rejecting. In order to give L an incentive to accept s_L but nothing less than that, M's decision must depend on the offer L rejected.[30] If L rejected an offer of s_L or more, then M punishes L by accepting $S(MR)$ from R. But if L rejected an offer of strictly less than s_L, then M rejects R's offer. If R's offer is rejected, L offers δs_M to M, and M accepts any offer of at least this much.

With these strategies, L clearly prefers to accept an offer of s_L. We also need to verify that he prefers to reject offers strictly below s_L. If L rejects an offer of $s_L - \varepsilon$, he must wait for two periods and then makes an offer of δs_M to M which M accepts. Thus, L rejects $s_L - \varepsilon$ for any $\varepsilon > 0$ if

$$s_L = S(LM) - s_M \leq \delta^2 \left[S(LM) - \delta s_M \right].$$

This inequality holds for δ close to 1 if $S(MR) > \frac{2}{3}S(LM)$ but is violated if $S(MR) < \frac{2}{3}S(LM)$ (using Equation (A.5)). Thus, Case β can only happen in the former case. M's payoff is given by Equation (A.5) and converges to $S(MR)$.

To summarize, if $S(MR) < \frac{2}{3}S(LM)$ then there is a Case α SSPE where $s_M \to \frac{2}{3}S(LM)$. If $S(MR) > \frac{2}{3}S(LM)$ then there is a Case β SSPE where $s_M \to S(MR)$.

(ii) If the protocol is LMR, then it is obvious that the LM coalition forms, since L is the formateur. The rest of the argument is similar to part (i), and is omitted. ∎

Stationary Equilibria with Random Recognition

In this section, we prove Proposition 2. Let ϕ_P denote the probability that player P is recognized in any period, and let s_P denote P's equilibrium expected payoff. By stationarity, s_P is P's expected continuation payoff at the beginning of any period.

Lemma A.4 *In any SSPE of the post-election bargaining game with repeated random recognition, $0 < s_P \leq S(LM)$ for each $P \in \{L, M\}$.*

Proof: It is obvious that $s_P \leq S(LM)$. Now the only way to get a negative payoff is to be left out of the government and suffer the externality. Player M is never left out because the LR government is ruled out by assumption.

---

[30] This does not contradict stationarity, since behavior does not depend on events in any previous round.

Moreover, with probability ϕ_M player M is recognized, and then he can offer δs_L to L which is accepted (because if L rejects he expects s_L next period). Therefore, M's continuation payoff is strictly positive,

$$s_M \geq \phi_M(S(LM) - \delta s_L) > 0.$$

It remains to show $s_L > 0$. To get a contradiction, suppose there is an SSPE with $s_L \leq 0$. Then M, if he is recognized, will certainly propose that he and L form a government where M gets all of the surplus $S(LM)$. (This is accepted because if L rejects he expects $s_L \leq 0$ next period.) If instead party $P \neq M$ is recognized, then M's payoff cannot be less than δs_M because he can always reject any offer and expect to get s_M next period; but neither will he be offered more than δs_M since offering δs_M makes M indifferent between accepting and rejecting. Thus, M gets δs_M when $P \neq M$ is recognized and $S(LM)$ when M is recognized, so M's equilibrium continuation payoff is

$$s_M = (\phi_L + \phi_R)\delta s_M + \phi_M S(LM).$$

Since $\phi_L + \phi_M + \phi_R = 1$, we get $s_M \to S(LM)$ as $\delta \to 1$. This means the MR government cannot form, since $S(MR) < S(LM)$, so L will never suffer the negative externality. With probability ϕ_L player L is recognized, and then he can offer $\delta s_M < S(LM)$ to M which is accepted. Therefore, L's continuation payoff is strictly positive, contradicting our hypothesis that $s_L \leq 0$.   ∎

Lemma A.5 *There is a unique SSPE in the post-election bargaining game with repeated random recognition. As $\delta \to 1$, the LM government always forms. If*

$$S(MR) < \frac{\phi_M}{\phi_L + \phi_M} S(LM) \tag{A.6}$$

then player M's share of the rents from office converges to

$$\frac{\phi_M}{\phi_L + \phi_M} S(LM). \tag{A.7}$$

If

$$S(MR) > \frac{\phi_M}{\phi_L + \phi_M} S(LM) \tag{A.8}$$

then player M's share of the rents from office converges to $S(MR)$.

Proof: Suppose in equilibrium R is *irrelevant* in the sense that any offer made by R is rejected and the equilibrium agreement is between L and M with probability 1. To see if this can be an equilibrium, we calculate the minimum amount that needs to be paid to induce acceptance by the player concerned (with the usual argument that in equilibrium each player accepts if indifferent). For $P \in \{L, M\}$, this minimum must be δ times the continuation payoff s_P which is positive by Lemma A.4. Thus, in each period, with probability ϕ_L player L makes an offer of δs_M to M, with probability ϕ_M player M makes an offer of δs_L to L, and with probability ϕ_R the game moves to the next period. Accordingly, the equilibrium payoffs must satisfy the following equations:

$$s_L = \phi_L(S(LM) - \delta s_M) + \phi_M \delta s_L + \phi_R \delta s_L,$$
$$s_M = \phi_L \delta s_M + \phi_M(S(LM) - \delta s_L) + \phi_R \delta s_M,$$
$$s_R = \phi_L(-x_R(LM)) + \phi_M(-x_R(LM) + \phi_R \delta s_R.$$

From the first two equations we get

$$s_M = \frac{\phi_M S(LM)}{1 - \delta + \delta\phi_M + \delta\phi_L}.$$

As $\delta \to 1$, $s_M \to \frac{\phi_M}{\phi_M + \phi_L} S(LM)$ as in Equation (A.7). If Condition (A.6) holds, then there is not enough surplus in the MR coalition for M to consider joining it, and R is truly irrelevant. But the equilibrium breaks down if Condition (A.8) holds, because R could offer M more than $\frac{\phi_M}{\phi_M + \phi_L} S(LM)$.

Now suppose Condition (A.8) holds, so R cannot be irrelevant in the above sense. We consider candidate stationary equilibria of three types. Type (i) is a pure strategy equilibrium with all offers accepted. Type (ii) involves mixed strategies in making offers. Type (iii) involves mixed strategies in response to offers. The stationarity requirement ensures that the continuation payoffs are the same in every period.

In type (i) equilibrium, M is offered δs_M by both L and R and accepts. If M is recognized, he offers δs_L to L. Therefore, a type (i) equilibrium gives us

$$s_L = \phi_L(S(LM) - \delta s_M) + \phi_M \delta s_L + \phi_R(-x_L(MR)),$$
$$s_M = \phi_L \delta s_M + \phi_M(S(LM) - \delta s_L) + \phi_R \delta s_M,$$
$$s_R = \phi_R(S(MR) - \delta s_M) + (1 - \phi_R)(-x_R(LM)).$$

As $\delta \to 1$, these expressions, when simplified, give us

$$s_M = \frac{\phi_L + \phi_R}{\phi_R} S(LM) + \phi_M \phi_R x_L(MR). \qquad (A.9)$$

But we know $s_M \leq S(LM)$, so Equation A.9 cannot hold. Therefore, a type (i) SSPE cannot exist.

Consider type (ii) equilibrium. Since L and R have only one option each, it must be M who randomizes (between offers to L and R). By an analysis similar to the above, we can show that this is impossible (the assumed probability with which M chooses L turns out to have a negative value). Therefore, a type (ii) SSPE cannot exist either.

The only possible stationary equilibrium therefore must be type (iii). Here again, it must be M who randomizes between accepting and rejecting an offer. This offer must be either $S(LM)$ from L or $S(MR)$ from R; for otherwise, the proposer could deviate and offer ε more, thus breaking the indifference and forcing M to accept with probability 1. But an offer of $S(LM)$ from L is clearly not part of any equilibrium (an offer of $\delta S(LM)$ would be acceptable since $s_M \leq S(LM)$). Thus, an equilibrium of type (iii) must have R offering all the surplus $S(MR)$ to M. Player M accepts this offer with probability y, where $0 < y < 1$.

In this (history independent) equilibrium, M must be indifferent between accepting and rejecting, so $\delta s_M = S(MR)$. Therefore M must make an acceptable offer to L, when it is his turn to do so, since $S(MR)/\delta$ is infeasible for the MR coalition. We can write down the equations as before, to describe the equilibrium expected payoffs. Once again R's payoffs are not important for the purpose of calculating the payoffs of the others:

$$s_L = \phi_L(S(LM) - S(MR)) + \phi_M \delta s_L + \phi_R[y(-x_L(MR)) + (1-y)\delta s_L],$$

$$s_M = \frac{S(MR)}{\delta} = \phi_L S(MR) + \phi_M(S(LM) - \delta s_L) + \phi_R S(MR).$$

The second equation gives

$$\delta^2 \phi_M s_L = \delta(1 - \phi_M)S(MR) - S(MR) + \delta \phi_M S(LM)$$

or

$$\delta s_L = \frac{\delta \phi_M S(LM) - (1 - \delta(1 - \phi_M))S(MR)}{\delta \phi_M}.$$

From these expressions, we can calculate the value of y as below:

$$(\delta^2 \phi_R s_L + \delta \phi_R x_L)y = \delta \phi_L[S(LM) - S(MR)] - \delta s_L(1 - \delta + \delta \phi_L). \qquad (A.10)$$

when $\delta = 1$, we have $s_L = S(LM) - S(MR)$, and substituting this in the expression above we get $y = 0$ when $\delta = 1$. We now wish to show that $y > 0$ for $\delta < 1$ but close to 1, so that the equilibrium exists for $\delta < 1$. We therefore differentiate both sides of the expression above with respect to δ and evaluate the derivative at $\delta = 1$. We first note that $\partial s_L / \partial \delta$ is positive at $\delta = 1$. Differentiating and setting $\delta = 1$, we get

$$2\phi_M s_L + \phi_M \frac{\partial s_L}{\partial \delta} = \phi_M[S(LM) - S(MR)] + S(MR)$$

or

$$\frac{\partial s_L}{\partial \delta} = \frac{S(MR)}{\phi_M} - [S(LM) - S(MR)]$$

$$> \frac{1}{\phi_M + \phi_L}(1 + \phi_M)S(LM) - S(LM) = \frac{S(LM)}{\phi_M + \phi_L}(1 - \phi_L) > 0$$

where the first inequality uses Condition (A.8). We now differentiate both sides of Equation (A.10) with respect to δ and evaluate again at $\delta = 1$, where we know $y = 0$. This gives us, using the above expression for $\partial s_L / \partial \delta$,

$$y\left(2\delta\phi_R s_L + \phi_R x_L + \delta^2\phi_R\frac{\partial s_L}{\partial \delta}\right) + (\delta^2\phi_R s_L + \delta\phi_R x_L)\frac{\partial y}{\partial \delta}$$

$$= \phi_L[S(LM) - S(MR)] - \delta s_L(-1 + \phi_L)$$

$$- s_L(1 - \delta + \delta\phi_L) - \delta(1 - \delta + \delta\phi_L)\frac{\partial s_L}{\partial \delta}$$

$$= (\phi_R s_L + \phi_R x_L)\frac{\partial y}{\partial \delta} = s_L(1 - \phi_L) - \phi_L\frac{\partial s_L}{\partial \delta}$$

$$= s_L(1 - \phi_L) - \phi_L\left(\frac{S(MR)}{\phi_M} - s_L\right)$$

$$= S(LM)[1 - \phi_L + \phi_L] - S(MR)\left[1 - \phi_L + \frac{\phi_L}{\phi_M} + \phi_L\right]$$

$$= S(LM) - S(MR)\frac{\phi_M + \phi_L}{\phi_M} < 0$$

by Condition (A.8). Therefore $\partial y/\partial \delta < 0$ at $\delta = 1$, so if δ decreases below 1, then y increases to $y > 0$, as was to be shown. ∎

Proposition 2 follows directly from Lemma A.5. It may be remarked that non-stationary pure strategy equilibria exist as well. Suppose Condition (A.8) holds and consider a non-stationary equilibrium of the following form. Player M offers δs_L to L, player L offers δs_M to M and player R offers $S(MR)$ to M, where

$$s_M = \phi_M(S(LM) - \delta s_L) + \phi_L \delta s_M + \phi_R S(MR)$$

so

$$s_M = \frac{\phi_M}{1 - \phi_L \delta}(S(LM) - \delta s_L) + \frac{\phi_R}{1 - \phi_L \delta}S(MR).$$

These offers are all accepted with probability 1. Player L's payoff is

$$s_L = \phi_L(S(LM) - \delta s_M) + \phi_M \delta s_L - \phi_R x_L(MR)$$

so

$$s_L = \frac{\phi_L}{(1 - \phi_M \delta)}(S(LM) - \delta s_M) - \frac{\phi_R}{(1 - \phi_M \delta)}x_L(MR).$$

These two equations can be solved for s_M and s_L. At $\delta = 1$, the expressions for s_M and s_L are as follows:

$$s_M = \phi_M[S(LM) + x_L(MR)] + (1 - \phi_M)S(MR) \qquad (A.11)$$

and

$$s_L = \phi_L[S(LM) - s_M] - \phi_R x_L(MR).$$

This is feasible if $s_L > 0$ (which requires $x_L(MR)$ small enough) and $s_M \leq S(LM)$. Here M can get strictly more than in the stationary equilibrium described earlier, provided he accepts R's offer (of less than his equilibrium payoff). This is therefore the only deviation we need to check, i.e., for M to reject R's offer. If he does, the continuation payoffs and the strategies are now given by the *stationary equilibrium above*, rather than by Equation (A.11). Thus, M's continuation payoff if he rejects R's offer is $S(MR)$, so he accepts the offer.

Ex Ante Coalitions Under Proportional Representation and ASB Protocol

In this section, we prove Propositions 3 and 4.

Case 1: In the absence of any ex ante agreement, L and M would form a coalition government ex post. In this case, there are three possible motives for forming the joint MR list ex ante: MR could form to give M its own majority, or to enable MR to form ex post, or to increase M's bargaining power via changing λ. We discuss these three possible motives in turn.

First motive. Suppose that the MR list results in M getting its own majority in parliament $(n(M) \geq 1/2)$. Then the government will be an M-party majority government (rather than an LM coalition). We need to check if it benefits both M and R. Now $n(M) \geq 1/2$ implies $n(R) \leq v(M)+v(R)-1/2$. That is, to achieve an M-party majority, R must give up at least $0.5-v(M)$ seats to M. This certainly makes M better off, and R gains $x_R(LM)-x_R(M)$ by blocking L from joining the government. The MR ex ante coalition is viable if R can be made better off, which holds if

$$x_R(LM) - x_R(M) > \alpha \left(0.5 - v(M)\right).$$

This is Condition (i) of Proposition 3. It requires that a coalition government involving L imposes a significant negative externality on R.

Second motive. Suppose the ex ante coalition between M and R impacts seat shares, and thus the ex post bargaining protocol, in such a way that the coalition government becomes MR rather than LM. By Proposition 1, this can only happen if M and R are not too ideologically distant, i.e., if $S(MR) > S(LM)/3$. If this inequality holds, then the MR coalition forms ex post if L is not the formateur and $\lambda = 1$. To accomplish this, R transfers seats to M via the joint list.[31] This transfer directly benefits M. Moreover, by shrinking, player R makes himself a more attractive coalition partner ex post.

Specifically, if $v(M) > v(R) > v(L)$, then with no ex ante agreement the protocol is MRL, and Proposition 1 implies that the LM government would form ex post. But if, through a joint list, R transfers $v(R) - v(L)$ seats to M, then R becomes the smallest party and the ex post bargaining protocol becomes MLR. Note that λ changes from 2 to 1 and, by Proposition 1, the coalition government changes to MR. Party R loses $v(R) - v(L)$ seats but now will be part of the coalition government, receiving a share $S(MR) - s^1(M)$ of the ex post surplus, and avoiding the externality $x_R(LM)$. Party M's share of the surplus falls from $s^2(M)$ to $s^1(M)$, but as compensation

---

[31] Here and in other places, ties can occur in seat shares, and the protocol must be defined even in this case. If R transfers seats to M ex ante in order to change the protocol to MLR, say, then if L and R end up with the same number of seats, we assume the protocol will actually be MLR. Otherwise, R would have to transfer ε more seats in order to make L strictly larger and so guarantee the order MLR. Since the possibility of ties is due to an artefact of the model, namely the continuum of seats, we assume this tie-breaking rule rather than making tedious ε arguments throughout.

he gains $v(R) - v(L)$ seats. Both M and R are made better off if

$$S(MR) - s^1(M) + x_R(LM) > \alpha\left(v(R) - v(L)\right) > s^2(M) - s^1(M).$$

This is Condition (ii) of Proposition 3.

But there are other possibilities. If $v(L) > v(M) > v(R)$ then the bargaining protocol without ex ante agreements is LMR. If through a joint list R transfers $v(L) - v(M)$ seats to M, then the ex post bargaining protocol becomes MLR. Again λ changes from 2 to 1 and the government changes from LM to MR. Both M and R are made better off if

$$S(MR) - s^1(M) + x_R(LM) > \alpha(v(L) - v(M)) > s^2(M) - s^1(M).$$

This is Condition (iii) of Proposition 3.

If instead $v(L) > v(R) > v(M)$ then the bargaining protocol without ex ante agreements is LRM. If via a joint list R transfers $v(L) - v(M)$ seats to M, then the ex post bargaining protocol becomes MLR. Here λ remains 1 but L is no longer the formateur, and by Proposition 1 the government changes from LM to MR. Player M is certainly better off because he gets more seats while his share of the surplus remains $s^1(M)$, and party R is better off if

$$S(MR) - s^1(M) + x_R(LM) > \alpha(v(R) - v(L)).$$

This is Condition (iv) of Proposition 3.

Third motive. The MR ex ante coalition might change the bargaining power within the LM coalition government in M's favor. If $\lambda = 1$, then R and M could both gain from an ex ante agreement where M transfers seats to R, so that λ changes from 1 to 2. For example, if $v(L) > v(R) > v(M)$, then with no ex ante agreement the protocol is LRM with $\lambda = 1$; but if M transfers $v(L) - v(R)$ seats to R, then R becomes the biggest party ex post and M remains the smallest, hence the ex post protocol is RLM, with $\lambda = 2$. This transfer of seats certainly makes R strictly better off, and M's share of the ex post surplus increases from $s^1(M)$ to $s^2(M)$ (as defined in Equation (1)) due to his increased bargaining power vis-a-vis L. Thus, M is strictly better off if $\lambda = 1$ and

$$s^2(M) - s^1(M) > \alpha\Delta$$

where Δ denotes the minimum number of seats that M needs to transfer to R to change λ from 1 to 2. This is Condition (v) of Proposition 3.

Case 2: In the absence of any ex ante agreement, M and R would form a coalition government ex post. Proposition 1 gives the conditions under which Case 2 occurs: $S(MR) > S(LM)/3$ and the bargaining protocol (in the absence of ex ante agreements) is either MLR or RML. Here, M and R cannot have a viable ex ante coalition (for the same reason that LM could not be viable in Case 1).

The ex ante coalition between L and M might be viable for several reasons. First, it might allow M to form a majority government. Such an ex ante agreement is viable if a coalition government which includes R has a big negative externality on L. The precise condition, analogous to Condition (i) of Proposition 3, is

$$x_L(MR) - x_L(M) > \alpha(0.5 - v(M)).$$

This is Condition (i) of Proposition 4.

The second way the ex ante coalition between L and M could be viable is if the ex ante agreement affects seat shares so the coalition government becomes LM rather than MR. This can be achieved in two ways: either by transferring seats from M to L, or by transferring seats from L to M. For example, if the bargaining protocol without ex ante agreements would be MLR, then L can transfer $v(L) - v(R)$ seat shares to M, making the new bargaining protocol MRL. Player M certainly gains from this. Party L loses seat shares, but now will be part of the coalition government, receiving a share $S(LM) - s^2(M)$ of the ex post surplus, and avoiding the externality $x_L(MR)$. Player L is better off if

$$S(LM) - s^2(M) + x_L(MR) > \alpha\left(v(M) - v(L)\right). \qquad (A.12)$$

But another way to change the coalition government from MR to LM is for M to transfer $(v(M) - v(L))/2$ seats to L, making the new bargaining protocol LMR (instead of MLR) Party L certainly gains from this. Party M loses seat shares, but gets a bigger share of the ex post surplus because λ has changed from 1 to 2. Player M is better off if

$$s^2(M) - s^1(M) > \alpha\frac{v(M) - v(L)}{2}. \qquad (A.13)$$

Analogous arguments can be made if the bargaining protocol without ex ante agreements would be RML. In either case, let Δ' denote the number of seats that L must transfer to M in order to change the coalition government from MR to LM, and let Δ'' denote the number of seats that M

must transfer to L in order to achieve the same outcome. Then we get the following two conditions, corresponding to Conditions (A.12) and (A.13), which are Conditions (ii) and (iii) of Proposition 4:

$$S(LM) - s^2(M) + x_L(MR) > \alpha\Delta'$$

and

$$s^2(M) - s^1(M) > \alpha\Delta''.$$

Ex Ante Coalitions under Plurality Voting and ASB Protocol

In this section, we consider in more detail the conditions for ex ante coalitions to be viable in Case 1 of the "ASB Bargaining ex post" section, where in the absence of ex ante agreement, L and M would form a coalition government.

Sub-case 1a: Party R is ex post irrelevant.

For the MR coalition to be viable ex ante, R must be made better off than if there is no ex ante coalition. By Proposition 1, in Case 1a, R will never be part of a coalition government. Moreover, recall that M cannot transfer seats to R under PV. The only way the ex ante agreement can make R better off is if it leads to a majority government formed by M (rather than an LM coalition government). For this to happen, R must drop out of some districts in order to raise M's seat share to $1/2$. If R drops out of y districts, then M will win a fraction $1 - z(L)$ of these, and a fraction $w(M)$ of the remaining $1 - y$ districts. Thus, y must satisfy

$$(1 - y)w(M) + y(1 - z(L)) \geq 0.5$$

and the smallest such y is

$$y = \frac{0.5 - w(M)}{1 - z(L) - w(M)}.$$

The cost to R of dropping out of y districts is $\alpha w(R)y$ (since he would win a fraction $w(R)$ in a three-way race). On the other hand, R gains $x_R(LM) - x_R(M)$ if M forms a one-party government rather than a coalition with L. Therefore, the condition for the MR ex ante coalition to be viable is that the negative externality an LM coalition government imposes on R is sufficiently big:

$$x_R(LM) - x_R(M) > \alpha w(R)\frac{0.5 - w(M)}{1 - z(L) - w(M)}. \tag{A.14}$$

This proves Proposition 7.

Sub-case 1b: Party R is ex post relevant.

As in Case 1a, the MR ex ante coalition is viable if Condition (A.14) holds, so the MR coalition enables party M to get its own majority. But in Case 1b, the MR coalition might be viable even if Condition (A.14) is violated. Specifically, because R is ex post relevant, forming the MR coalition ex ante might change the coalition government from LM to MR. To avoid tedious repetition, assume to be specific that $w(M) > w(R) > w(L)$. (The other cases can be worked out similarly.) If there is no ex ante agreement, then $\lambda = 2$ so the LM coalition forms ex post. The MR coalition government instead form if L is not the formateur and $\lambda = 1$. This can be achieved by making sure that R is the smallest party, so the protocol becomes MLR (with $\lambda = 1$) instead of MRL. Let R drop out of r seats. Then M gets $(1 - r)w(M) + r(1 - z(L))$ seats, L gets $(1 - r)w(L) + rz(L)$, and R gets $(1 - r)w(R)$. We require L to get more seats than R, that is,

$$(1 - r)w(L) + rz(L) \geq (1 - r)w(R)$$

which is the same as

$$r \geq \frac{w(R) - w(L)}{w(R) - w(L) + z(L)}. \tag{A.15}$$

For the MR government to form, M must have less than half of the seats,

$$(1 - r)w(M) + r(1 - z(L)) \leq \frac{1}{2}$$

which is the same as

$$r \leq \frac{\frac{1}{2} - w(M)}{1 - z(L) - w(M)}. \tag{A.16}$$

Player R's payoff will be $\alpha(1 - r)w(R)$. For R to be willing to participate, we need

$$\alpha(1 - r)w(R) > \alpha w(R) - x_R(LM)$$

which is the same as

$$r < \frac{x_R(LM)}{\alpha w(R)}. \tag{A.17}$$

Thus, the MR ex ante coalition is viable if there exists r satisfying Conditions (A.15), (A.16) and (A.17).

References

Austen-Smith, D. and J. Banks. 1988. "Elections, Coalitions, and Legislative Outcomes." *American Political Science Review* 82: 405–422.

Axelrod, R. 1970. *The Conflict of Interest.* Chicago, IL: Markham.

Bandyopadhyay, S. and K. Chatterjee. 2006. "Coalition Theory and its Applications: A Survey." *Economic Journal* 116: 136–155.

Bandyopadhyay, S. and M. Oak. 2004. "Party Formation and Coalitional Bargaining in a Model of Proportional Representation." Discussion Paper. University of Birmingham, UK.

Bandyopadhyay, S. and M. Oak. 2008. "Coalition Governments in a Model of Parliamentary Democracy." *The European Journal of Political Economy* 24: 554–556.

Baron, D. and J. Ferejohn. 1989. "Bargaining in Legislatures." *American Political Science Review* 83: 1181–1206.

Baron, D. and D. Diermeier. 2001. "Elections, Governments, and Parliaments under Proportional Representation." *Quarterly Journal of Economics* 116: 933–967.

Baron, D., D. Diermeier, and P. Fong. 2011. "A Dynamic Theory of Parliamentary Democracy." *Economic Theory* (forthcoming).

Battaglini, M. and S. Coate. 2007. "Inefficiency in Legislative Policymaking: A Dynamic Analysis." *American Economic Review* 97: 118–149.

Battaglini, M. and S. Coate. 2008. "A Dynamic Theory of Public Spending, Taxation, and Debt." *American Economic Review* 98: 201–236.

Besley, T. and S. Coate. 1997. "An Economic Model of Representative Democracy." *Quarterly Journal of Economics* 112: 85–106.

Blais, A. and I. Indridason. 2007. "Making Candidates Count: The Logic of Electoral Alliances in Two-Round Legislative Elections." *The Journal of Politics* 69: 193–205.

Brunell, T. and B. Grofman. 2009. "Testing Sincere versus Strategic Split-Ticket Voting at the Aggregate Level: Evidence from Split House–President Outcomes, 1900–2004." *Electoral Studies* 28: 62–69.

Carroll, R. and G. Cox. 2007. "The Logic of Gamson's Law: Pre-election Coalitions and Portfolio Allocations." *American Journal of Political Science* 51: 300–313.

Chatterjee, K., B. Dutta, D. Ray, and K. Sengupta. 1993. "A Noncooperative Theory of Coalitional Bargaining." *Review of Economic Studies* 60: 463–477.

Debus, M. 2009. "Pre-electoral Commitment and Government Formation." *Public Choice* 138: 45–64.

Degan, A. and A. Merlo. 2007. "Do Voters Vote Sincerely?" Working Paper, CEPR.

Dhillon, A. 2005. " Political Parties and Coalition Formation." In *Networks, Clubs and Coalitions,* Demange, G. and M. Wooders, eds., Cambridge, UK: Cambridge University Press, pp. 289–311.

Diermeier, D. and A. Merlo. 2000. "Government Turnover in Parliamentary Democracies." *Journal of Economic Theory* 94: 46–79.

Diermeier, D. and A. Merlo. 2004. "An Empirical Investigation of Coalitional Bargaining Procedures." *Journal of Public Economics* 88: 783–797.

Downs, A. 1957. *An Economic Theory of Democracy.* New York, NY: Harper.

Eraslan, H. and A. Merlo. 2002. "Majority Rule in a Stochastic Model of Bargaining." *Journal of Economic Theory* 103: 31–48.

Fong, P. 2006. "Dynamics of Government and Policy Choice." Working Paper, Concordia University.

Gallagher, M., M. Laver, and P. Mair. 1995. *Representative Government in Modern Europe.* New York, NY: McGraw Hill.

Golder, S. 2006a. "Pre-electoral Coalition Formation in Parliamentary Democracies." *British Journal of Political Science* 36: 193–212.

Golder, S. 2006b. *The Logic of Pre-Electoral Coalition Formation.* Columbus, OH: Ohio State University Press.

Golder, S. and J. Thomas. 2011. "Gamson's Law? Portfolio Allocation and the Vote of No-Confidence." Working Paper, Pennsylvania State University.

Indridason, I. 2003. "Coalitions and Clientelism: A Comparative Study." Working Paper, Department of Political Science, University of Iceland.

Indridason, I. 2005. "A Theory of Coalitions and Clientelism: Coalition Politics in Iceland, 1945–2000." *European Journal of Political Research* 44: 439–464.

Indridason, I. 2009. "Proportional Representation, Majoritarian Legislatures and Coalitional Voting." Working Paper, University of California, Riverside.

Indridason, I. 2010. "Live for Today, Hope for Tomorrow? Rethinking Gamson's Law." Working Paper, University of California, Riverside.

Indrdason, I. 2011. "Coalition Formation and Polarization." *European Journal of Political Research* (forthcoming).

Jackson, M. and B. Moselle. 2002. "Coalition and Party formation in a Legislative Voting Game." *Journal of Economic Theory* 103: 49–87.

Laver, M., S. de Marchi, and H. Mutlu. 2010. "Negotiation in Legislatures over Government Formation." *Public Choice* (forthcoming).

Laver, M. and N. Schofield. 1990. *Multiparty Governments: The Politics of Coalitions in Europe*. Oxford, UK: Oxford University Press.

Laver, M. and K. Shepsle. 1996. "Making and Breaking Governments: Cabinets and Legislatures in Parliamentary Democracies." New York, NY: Cambridge University Press.

Levy, G. 2004. "A Model of Political Parties." *Journal of Economic Theory* 115: 250–277.

Martin, L. and R. Stevenson. 2001. "Government Formation in Parliamentary Democracies." *American Journal of Political Science* 45: 33–50.

Morelli, M. 2004. "Party Formation and Policy Outcomes under Different Electoral Systems." *Review of Economic Studies* 71: 829–853.

Okada, A. 1996. "A Noncooperative Coalitional Bargaining Game with Random Proposers." *Games and Economic Behavior* 16: 97–108.

Okada, A. 2007. "Coalitional Bargaining Games with Random Proposers: Theory and Applications." Working Paper, Hitotsubashi University Graduate School of Economics, Tokyo, Japan.

Osborne, M. and A. Slivinski. 1996. "A Model of Political Competition with Citizen-Candidates." *Quarterly Journal of Economics* 111: 65–96.

Osborne, M. and R. Tourky 2002. "Party Formation in Collective Decision-Making." Working Paper, The University of Melbourne, Australia.

Pech, G. 2010. "A Median Voter Theorem for Proportional Representation Systems with Firm Pre-electoral Coalitions." Working paper, KIMEP.

Persson, T., G. Roland, and G. Tabellini. 2007. "Electoral Rules and Government Spending in Parliamentary Democracies." *Quarterly Journal of Political Science* 2: 155–188.

Powell, B. G. 2000. *Elections as Instruments of Democracy*. New Haven, CT: Yale University Press.

Pugh, M. 2002. *The Making of Modern British Politics, 1867–1945*. Blackwell.

Ray, D. 2008. *A Game-Theoretic Perspective on Coalition Formation*. Oxford, UK: Oxford University Press.

Riker, W. 1962. *The Theory of Political Coalitions*. New Haven, CT: Yale University Press.

Roemer, J. 2001. *Political Competition: Theory and Applications*. Cambridge, MA: Harvard University Press.

Snyder, J. and M. Ting. 2002. "An Informational Rationale for Political Parties." *American Journal of Political Science* 46: 90–110.

Strom, K., W. Mueller, and T. Bergman. 2008. *Cabinets and Coalition Bargaining; The Democratic Life Cycle in Western Europe*. Oxford, UK: Oxford University Press.

IEEE TRANSACTIONS ON SYSTEMS, MAN, AND CYBERNETICS, VOL. SMC-11, NO. 2, FEBRUARY 1981

101

Comparison of Arbitration Procedures: Models with Complete and Incomplete Information

KALYAN CHATTERJEE

Abstract— This paper has focused on two types of arbitration procedures that are commonly used in the United States, especially in industrial disputes involving public safety personnel who are legally not permitted to strike. The conventional arbitration (CA) procedure gives the arbiter or arbitration panel complete authority to fashion a final award. The final-offer arbitration (FOA) procedure allows the arbiter only to choose one of the two final offers made by the parties. Though some empirical work has appeared comparing the two procedures, the theoretical analysis of the procedures is still incomplete. This paper has concentrated on the effect of uncertainty about the arbiter's preferences on the bargainers, while ignoring other aspects of the comparison such as permitting the arbiter to trade-off among bargaining issues.

I. Introduction

COMPULSORY binding interest arbitration is becoming an increasingly accepted form of third-party intervention in industrial disputes where work stoppages are socially undesirable, such as those involving public safety employees. Seventeen states now have statutory provisions for an arbitration recourse in the event of a bargaining impasse. In some cases arbitration has supplanted other modes of third-party intervention like mediation and fact-finding with nonbinding recommendations; in many others it has supplemented them as a last resort.

The arbitration procedures in different states differ in several important aspects (see [7]–[9], [13], and [14]). We shall, however, consider only two categories of procedure: conventional arbitration (CA) and final-offer arbitration (FOA).

In *conventional arbitration* the arbiter is unrestricted in his freedom to fashion what he might consider a reasonable award and to impose it on the participants. Though the statutes usually mention some criteria for making the award, the arbiter usually has a great deal of leeway in choosing a solution.

This has led to the criticism that the arbitration process allows third parties ("itinerant philosophers" as arbiters are sometimes unflatteringly described) to impose their values on the bargainers. Typically, employers have complained of a bias in favor of employees leading to unreasonably high awards on wages and other economic issues.

A second criticism of conventional arbitration has been that it leads to a "chilling" effect on the bargaining. The availability of arbitration, critics say, relieves the bargainers of the sometimes politically unpleasant task of compromising, and the collective bargaining process loses its significance.

Final-offer arbitration was developed to meet these criticisms. In this mode of arbitration the parties bargain. If they fail to reach an agreement (even after intermediate steps like mediation and fact-finding), the dispute is sent to the arbiter who asks each party to submit a final offer. The arbiter is then restricted to choosing one of the two final offers as the final settlement.

Proponents of final-offer arbitration argue that the procedure greatly reduces the chilling effect of arbitration on bargaining, since each side will moderate its offers in order to appear reasonable to the arbiter. It would also, in their view, diminish the amount of arbitrariness in arbitration, since the arbiter would have to choose a solution reflecting the values and preferences of (at least one of) the participants.

Empirical findings have not settled the question one way or the other. Earlier theoretical work [4]–[6] has arrived at the conclusion that there is no substantive difference between the results of FOA and conventional arbitration. Our theoretical models, whose rationale is described more fully in the next section, do not support this result. In our first model, which most closely resembles previous work, we show that the introduction of uncertainty about the arbiter's choice makes bargainers adopt more extreme positions under FOA than under conventional arbitration, a result that would appear to undermine the FOA argument at what appears its strongest point. In other models we examine the effects of different information patterns on the procedures and relate arbitration to our earlier models of bargaining described in [1]–[3]. As a continuation of the research described here, one direction would be to try to relate the theoretical approaches to the actual case histories of arbitration. We do not attempt this here.

Manuscript received June 25, 1980; revised October 17, 1980. This work was supported in part by the Defense Advanced Research Projects Agency through their sponsorship of the Program of Advanced Research on Temporal and Group Decision Making at Harvard University under contract N0014-76-0074.

The author is with the Division of Management Science, Pennsylvania State University, 310 Business Administration Building, University Park, PA 16802.

II. Preliminaries to Formal Modeling

In this section we discuss the various issues that need to be considered in a theoretical comparison of the two procedures and offer a rationale for the questions that we do consider in the theoretical models of the following sections.

A. Single-Issue Versus Multiple-Issue Disputes

We are going to limit ourselves to disputes along a single dimension, for example, a dispute on the wage rate. This has the disadvantage, as has been pointed out [4], that all bargaining outcomes are ex post Pareto optimal, so that issues relating to the ex post efficiency of FOA and conventional arbitration are not considered. However the assumption of single-issue bargaining serves to keep our models analytically tractable as we introduce uncertainty and differential information. We are able to obtain insights with our single-issue models on the important question of the effect of the arbitration option on collective bargaining, in particular on the magnitude of the "chilling" effect, if any.

B. How Many Stages Should Our Models Have

As mentioned earlier the process of resolution of industrial disputes could (and does) consist of many stages — beginning with bargaining and encompassing mediation, fact-finding, the "final offers" handed in at the beginning of arbitration hearings, and the amended final offers at the end of the hearings. We shall restrict ourselves to two stages: a bargaining stage and an arbitration stage. We implicitly include fact-finding with recommendations as well as those situations where we assume that the parties to the dispute know what the arbiter's ideal solution is. Though the arbiter and the fact finder are usually different persons, empirical studies [8] have noted that arbitration awards are exactly the same as the fact finder's recommendations in 70 percent of the cases studied.

In the first stage of the process, the parties bargain with each other. If they agree, the process terminates. If they fail to agree, the second stage begins: the arbitration stage. At the end of arbitration, the process terminates. (The bargaining–arbitration sequence is repeated every time a new contract is being negotiated. We consider only a single negotiation and do not discuss possible interactions between negotiations.)

C. Criteria for the Arbiter's Choice

Broadly speaking, the arbiter's choice could depend upon the offers made by the bargainers and his or her own equity judgments. We distinguish between those arbitration procedures that are independent of the bargainers' offers and those that are so dependent.

Conventional arbitration, at least of the type that is considered in this paper, depends only on the arbiter's perceptions of equity. The arbitration award is then just the arbiter's equity-maximizing solution. An alternative form of conventional arbitration, one that is not discussed here, could for example, have the arbiter choosing a solution midway between the offers of the two sides and thus make the award responsive to the offers made by the bargainers.

Final-offer arbitration would, of course, always depend on at least one of the offers submitted. We discuss two kinds of final-offer arbitration here, one that depends solely upon the closeness of the two offers to the arbiter's equity-maximizing solution, and one that seeks to reward bargainers for some compromise relative to known extreme initial positions taken by the two sides. Professional arbiters apparently often dislike the second approach because it leads to gamesmanship during the bargaining process.

A third type of FOA that we do not consider here would have the arbiter choosing each offer with probability $1/2$ as the equivalent of the conventional arbitration approach of choosing the middle point between two offers. (It is clear of course that risk-averse individuals would prefer the conventional approach, once they were in the bargaining phase.)

D. The Basic Bargaining Model and Extension to Arbitration

As mentioned earlier we are not going to be concerned with multiattribute bargaining situations in this paper, so we shall not look into the distinctions between selection of package final offers and selection of final offers issue by issue. We shall consider bargaining along one dimension, e.g., the wage rate, with the implicit understanding that all other issues have either been settled or are not up for negotiation.

We further assume that the range of possible bargaining outcomes is known to both parties and to the arbitrator. This range, for example, would be given by the current wage at the lower end and by the management's maximum ability to pay at the upper end. The latter quantity is probably not initially known but could be determined by the parties by asking management to provide evidence to back any claim it makes (such as access to financial statements). This is, of course, not a complete description of labor–management bargaining, especially in cases where strikes are permitted. Labor might be willing to strike rather than to accept the current wage, and management might prefer no agreement to one that stretches its resources to the limit. In this case the true range would be smaller than the one we have assumed to be the case. Our rationale for this is that strikes are not permitted in the environment in which our bargainers work. The arbitration recourse is our alternative to strikes, and we do model explicitly the effect of arbitration on the range of agreements that can be concluded in the bargaining phase. We do not deal, in this paper, with the situation where the end points of the bargaining range (say, the current wage and management's maximum ability to pay) are either both unknown or partially unknown. We normalize outcomes so that the bargaining range is [0, 1] where the maximum ability to pay of management is denoted by 1.

The most realistic formulation of the bargaining would be as a noncooperative multistage game. However such a game is difficult to solve analytically. We therefore consider one-stage approximations for all the analytical results. These one-stage models bear a close family resemblance to earlier work by the author and others on bargaining under uncertainty ([1]–[3], [11], and [12]). In one of these earlier bargaining models we consider a situation where each of the two bargaining parties has a *reservation price* that embodies the maximum limit of concessions the bargainer is prepared to make. The exchange of offers (in our model, a simultaneous exchange) takes place against the backdrop of the underlying reservation price structure that may be partially unknown to one or both of the players. The special feature in this paper is that the reservation price for the bargaining phase is set by the expected outcome in the arbitration phase. In FOA or in any type of arbitration procedure responsive to the offers made by the bargaining parties, the expected payoff in the arbitration phase depends on the actions of players in that phase and is thus controllable to some extent. If the actions of players in the bargaining phase also affect their payoff in the arbitration phase we have a complicated interdependence which determines both the reservation prices and the actual outcome.

It might be interesting to pause for a moment to consider the relationship of our arbitration models with the arbitration model of Raiffa and the bargaining model of Nash (see [10]). If the Nash/Raiffa solutions are interpreted as arbitration procedures the work of the arbiter is conceived as the determination of an equitable efficient outcome in the range of bargaining alternatives. In the event of conflict the arbiter has no role. In our models the arbiter essentially specifies a *conflict* payoff, which is relevant only when there is no agreement. If an agreement is possible the bargainers divide the surplus equally.

This difference is related to the point of view we adopt that bargaining is essentially a noncooperative process with players having to decide whether to hold out for individual gain or to go along with an efficient cooperative solution (see [1]–[3] for further details). Thus, in our models, it is entirely possible for an impasse to occur, even if a mutually beneficial agreement exists. This is most clearly expressed in Section IV on models with differential information. In our other models as well, impasses could occur if, for example, labor (hereafter referred to as player L) were to demand more than management (player M) were willing to accept. We therefore adopt the noncooperative solution concept of Nash equilibrium in all cases (see [10] for a discussion of this concept, including the Harsanyi extension to Bayesian Nash equilibrium in Section IV (see [11] and [12] for further discussion). A detailed discussion of various arbitration models in relation to their use of cooperative and noncooperative concepts is contained in Crawford [5].

E. Informational Assumptions

We shall assume throughout that the rules of the game, including the procedure by which an arbiter arrives at a final choice, are known to everybody. Other aspects of the process may be completely known to all parties or completely known to some and partially to others or incompletely known.

For example, we could assume *complete information*: both parties know all data of the problem, including the arbitrator's equity-maximizing solution (though each does not know what the other is going to do); each party also knows any possible costs borne by itself and its adversary as a result of continuing bargaining as well as of going to arbitration.

A less stringent assumption would be *incomplete but identical information*. That is, both parties would be uncertain about the arbitrator's ideal solution but would have the same information (embodied in a probability distribution) about what it could be. We could compare modes of behavior under this assumption to those under the complete information assumption and check whether it is better for both parties (during bargaining) for the arbitrator to reveal what an externally desirable solution might be.

Finally we could look at models with *differential information* in which each side has private information not available to the other party or to the arbitrator. One way to include differential information is to assume that bargainers have different stage costs or time costs. This would appear to give these costs a significant role in determining bargaining posture, as does happen with the last model we consider. It is not clear how large these costs could be in comparison to the settlement value. It could be substantial if a bargainer has limited resources and has to hire professional negotiators during arbitration. In addition, however, the implicit or perhaps psychic costs in going from the bargaining to the arbitration stage might explain why the industrial relations literature places so much stress on the outcome being determined in the bargaining phase. This brings us to the next issue in modeling the process.

F. Criteria for Evaluating Alternative Procedures

Most of the literature in the labor relations area seems to judge the success of an arbitration scheme by its deterrent effect. In other words, how successful has the scheme under consideration been in persuading bargainers to settle all issues without the aid of the arbitration mechanism; that is, how has a scheme affected the probability of agreement in the bargaining phase? This seems to be too narrow to use as the sole criterion to evaluate a procedure. If the arbitrator informs bargainers about his ideal solution this might well have an effect on getting the bargainers to argue in the neighborhood of the solution and to reach an agreement thereby. However the solution is imposed and may not satisfy symmetry and fairness conditions natural to the problem. It could also restrict the range of choices available to the bargainers. In the case where there is more

104 IEEE TRANSACTIONS ON SYSTEMS, MAN, AND CYBERNETICS, VOL. SMC-11, NO. 2, FEBRUARY 1981

than one issue at stake, an imposed solution is almost certainly inefficient. In our case the efficiency issue is sidestepped by the special nature of the "distributive bargaining" game we are considering.

III. ONE-STAGE MODELS UNDER COMPLETE OR INCOMPLETE BUT IDENTICAL INFORMATION

Complete Information: Here we shall assume that three arbitration schemes, two of which have a possible interpretation as FOA, exist.

First, the arbitrator imposes a solution x which represents his ethical and other details. The solution x is known to both parties. We shall call this conventional arbitration (CA).

Second, the arbitrator chooses the offer (in the arbitration phase) which is nearest x in terms of absolute value. We shall refer to this as FOA 1.

Third, the arbitrator selects the offer which represents the less extreme demand (or the greater concession from one or management, respectively, for labor and management). We call this FOA 2.[1]

We assume that the offers used by the arbiter are the same final offers the parties made to each other during bargaining. We assume that the costs of going from bargaining to arbitration and the costs incurred during bargaining are all zero. (This assumption will be relaxed later.)

Now consider the game that results. Formally, the game is as follows. Players L and M submit offers a_l and a_m, respectively, to each other. If $a_m \geqslant a_l$, there is an agreement at $(a_m + a_l)/2$. If not, the arbiter is called in and he or she uses either CA, FOA 1, or FOA 2.

If x is the arbitrated outcome, the payoffs are

$$1 - x \quad \text{to player } M \tag{1}$$

and

$$x \quad \text{to player } L. \tag{2}$$

If $a_m \geqslant a_l$, the payoffs are

$$1 - \frac{a_m + a_l}{2} \quad \text{to player } M \tag{3}$$

and

$$\frac{a_m + a_l}{2} \quad \text{to player } L. \tag{4}$$

Now it is clear that the following holds.

Proposition 1: In the game as described above, a Nash equilibrium strategy pair is for each to offer

$$a_l = a_m = x \tag{5}$$

if CA or FOA 1 is used as the arbitration procedure.

[1] Mathematically, of course, FOA 2 is a special case of FOA 1 with $x = 1/2$. However, FOA 1 and FOA 2 imply different kinds of arbitration behavior: one considering only the closeness of the final offers to the arbiter's preferences, the other relating the final offers to the initial offers and noting the extent of the concessions made.

Proof: The proof is obvious (see [4] for a much more general proof).

However, if FOA 2 is used, the focal point is different. Before starting the equilibrium analysis, it might be advisable to state the procedure formally. Suppose player L offers $a_l \leqslant a_m$, there is an agreement at $(a_l + a_m)/2$. If $a_l > a_m$, the arbitrator chooses the minimum of $(1 - a_m, a_l)$. That is, he chooses a_m, if $1 - a_m < a_l$ and vice versa for the opposite strict inequality. In the case where $1 - a_m = a_l$, the arbitrator tosses a coin to choose between a_l and a_m.

Proposition 2: Under FOA 2, a pure strategy Nash equilibrium is for players to make their offers such that

$$a_l = a_m = \frac{1}{2}. \tag{6}$$

Proof: Suppose $a_m = 1/2$. Player L will not offer anything less than $1/2$, since he could gain by moving to $1/2$.

However if player L offers $a_l > 1/2$, the arbitrated solution will be a_m, so that he cannot gain by offering more than $1/2$. A similar chain of reasoning holds for player M (this result would be identical, if FOA 1 is used and $x = 1/2$).

If player L makes an offer $a_l > 1/2$, say $3/4$, what is player M's best response offer? Clearly, $a_m \leqslant 3/4$, but there is no reason why player M should stop below $a_m = 1/4 + \epsilon$. If he offers this, he is certain to win the award and obtain a payoff $(3/4 - \epsilon)$ where ϵ can be made arbitrarily small. Therefore, points on the line apart from $1/2$ are not equilibria. This is different from the case of CA where (with zero costs) any offer however extreme would lead to the same solution. FOA 1 gives a similar unique equilibrium.

The arbitration games considered above set reservation prices for the bargaining phase. For example, if player L knows that CA or FOA 1 is to be used if there is no agreement, he would not settle for less than x in the bargaining. Player M, on the other hand, would not be willing to accept any compromise greater than x. This effectively limits the bargaining to one point. A similar argument holds for FOA 2.

We have assumed costs of going to the arbitration phase to be zero. If we introduce some costs, c_l and c_m to labor and management, respectively, the reservation prices for CA and FOA 1 become $x - c_l$ for player L and $x + c_m$ for player M. If the costs are both known we could predict (maybe not very successfully) the bargaining outcome by using something like the Nash scheme. The case where costs are not known is covered later.

Incomplete Information: Now suppose that the arbitrator's ideal solution x is unknown to the players but that they share common probabilistic assessments (summarized in a distribution $F(\cdot)$) on x. Suppose further that both players are risk-neutral.

Proposition 3: Under conventional arbitration, the players will offer, in equilibrium,

$$a_l = a_m = \bar{x}. \tag{7}$$

Proof: Obvious.[2]

The situation is somewhat different under FOA 1. Assume that both players are risk-neutral again. Consider the arbitration phase as a separate game as before. Assuming the player M is going to play a_m, what would be player L's best response in pure strategies? He could either choose $a_l = a_m$ or decide to choose $a_l > a_m$ and maximize his expected value given the probability distribution on the arbitrator's award.

The arbitrator's choices are as follows.

If $x > a_l > a_m \to$ choose a_l.

If $a_l > x > a_m$

and

$$a_l - x < x - a_m \to \text{choose } a_l.$$

If $a_l > x > a_m$

and

$$a_l - x - a_m \to \text{choose } a_m.$$

If $a_l > a_m > x \to$ choose a_m.

Whenever the equality sign holds, randomization is used. Player L's expected value under a distribution $f(\cdot)$ is

$$\int_{(a_l + a_m)/2}^{\infty} a_l f(x) \, dx + \int_{-\infty}^{(a_l + a_m)/2} a_m f(x) \, dx. \quad (8)$$

Notice that (8) minus a_m is always positive, so that it is always to player L's advantage to state $a_l > a_m$. Differentiating this and by setting it equal to zero to obtain the optimal a_l in the interior of the region $[a_m, 1]$, we get

$$1 - F\left(\frac{a_l + a_m}{2}\right) - \frac{a_l}{2} f\left(\frac{a_l + a_m}{2}\right) + \frac{a_m}{2} f\left(\frac{a_l + a_m}{2}\right) = 0. \quad (9)$$

If $F(\cdot)$ is uniform on $[\alpha, \beta]$, this simplifies to

$$\left(\beta - \frac{a_l + a_m}{2}\right) + \frac{a_m}{2} - \frac{a_l}{2} = 0, \quad (10)$$

or

$$(\beta - a_l) = 0. \quad (11)$$

This gives $a_l = \beta$, irrespective of a_m. Thus we have the following proposition.

Proposition 4: Under FOA 1, and with $\bar{x}$ uniform, a dominant strategy for labor is to make the most extreme demand feasible.

Therefore, under certain conditions, FOA 1 could lead to more extreme positions than CA.[3] This could happen by

one player reasoning that if the other also took an extreme position, each had a good chance of winning a large amount, while if the other player took a more moderate position, the arbitration award would be lost but it would not matter because it would be more acceptable anyway.

The rather surprising result obtained for the uniform distribution does not generalize. That is, it is not true in general that player L's best strategy is to announce β irrespective of player M's choice a_m. However, even in other cases it seems a player would gain by extreme demands. We consider just one case.

Let x be distributed on $[0, \beta]$ with a density function

$$f(x) = \frac{2}{\beta} - \frac{2x}{\beta^2}.$$

Notice that this is largest when $x = 0$. We would expect therefore that player L, fearing a low value of x, would moderate his offer. However, applying (9) above we obtain

$$1 - \frac{2}{\beta}\left(\frac{a_l + a_m}{2}\right) + \frac{1}{\beta^2}\left(\frac{a_l + a_m}{2}\right)$$

$$+ \frac{(a_m - a_l)}{2}\left(\frac{2}{\beta} - \frac{2}{\beta^2} \cdots \frac{a_l + a_m}{2}\right) = 0,$$

or

$$1 - \frac{a + a_m}{\beta} + \frac{1}{4\beta^2}\left(a_l^2 + a_m^2 + 2a_1 a_m\right)$$

$$+ \frac{2}{\beta}\left(\frac{a_m - a_l}{2}\right) \cdot \left(1 - \frac{1}{2}(a_1 - a_m)\right) = 0.$$

Simplifying, we get

$$1 - \frac{2a_l}{\beta} + \frac{3}{4}\frac{a_l^2}{\beta^2} - \frac{a_m^2}{4\beta^2} + \frac{a_l a_m}{2\beta^2} = 0.$$

Remembering that $\beta \geqslant a_l > a_m$, we see that the derivative of the above expression is negative so that the second-order condition is satisfied. Further, a quick calculation shows that

$$\frac{da_l}{da_m} \geqslant 0.$$

Thus, the offers move in the same direction. Now suppose that $a_m = 0$, the most extreme management offer. Substituting above we get

$$a_l = \frac{2}{3}\beta.$$

However, as management becomes more moderate, it is to labor's advantage to become more extreme, so that the procedure appears to contain incentives for doing exactly the thing its proponents wish to avoid.

IV. A MODEL WITH DIFFERENTIAL INFORMATION

A. A Description of FOA and Conventional Arbitration in a Simple Model with Differential Information

Conventional Arbitration: Two players, labor (player L) and management (player M), bargain on a wage bill which

---

[2] However, if players L and M have different utility functions, the lottery on $\bar{x}$ will be evaluated differently by them. The arbitration game would be worth certainty equivalent (CE)(M) to player L and (CE)(L) to management. These would then be their reservation prices for the bargaining phase. If one player's CE is not known to the other, we are in the situation where players have private information about their own reservation prices and bargain noncooperatively, using their information to gain strategic advantage. This is analyzed elsewhere.

[3] As pointed out by a referee, the equilibrium of Proposition 3 is only one equilibrium, and more extreme positions are also in equilibrium (because it does not matter what the bargainers do, so long as they do not agree). This might appear to undermine the position taken here that FOA leads to more extreme positions than CA. However, the incentive to shift to extreme positions in CA is clearly a weak one (you do not lose by so doing), while in FOA, bargainers can expect to gain by extreme positions.

106   IEEE TRANSACTIONS ON SYSTEMS, MAN, AND CYBERNETICS, VOL. SMC-11, NO. 2, FEBRUARY 1981

has to be in the range $[0, 1]$ where the current wage bill is normalized at zero, and the known maximum profit level (or maximum ability to pay of the management) at one.

The game proceeds as follows: there are independent random drawings from distributions $F_l(\cdot)$ and $F_m(\cdot)$. Player L is told the value of c_l, the result of the first drawing, and player M the value of c_m, the second drawing. In the next move the players simultaneously make offers. Let player L's offer be denoted by a_l and player M's offer by a_m, where

$$a_l = A_l(c_l)$$

and

$$a_m = A_m(c_m)$$

where the $A_l(\cdot)$ and $A_m(\cdot)$ functions relate a player's act to his or her information. If $a_m \geqslant a_l$, the payoffs are

$$\frac{a_m + a_l}{2} \qquad \text{to player } L \qquad (12)$$

and

$$1 - \frac{a_m + a_l}{2} \qquad \text{to player } M.$$

If $a_m < a_l$, the arbitrator imposes a solution x and the payoffs are

$$x - c_l \qquad \text{to player } L \qquad (13)$$

and

$$1 - x - c_m \qquad \text{to player } M.$$

The returns R_l^c, R_m^c to each player can be written as

$$R_l^c(a_l, a_m; c_l) = \left(\frac{a_m + a_l}{2}\right)\delta(a_m, a_l) + (x - c_l)(1 - \delta(a_m, a_l)), \qquad (14)$$

$$R_m^c(a_l, a_m; c_m) = \left(1 - \frac{a_m + a_l}{2}\right)\delta(a_m, a_l) + (1 - x - c_m)(1 - \delta(a_m, a_l)), \qquad (15)$$

where

$$\delta(a_m, a_l) = 1, \qquad a_m \geqslant a_l,$$
$$= 0, \qquad a_m \leqslant a_l. \qquad (16)$$

In other words, the failure of bargaining implies costs to both parties and each party's cost is unknown to the other. The arbitrator's solution is externally decided and imposed.

Final-Offer Arbitration: The informational aspects of the model are the same as in the conventional arbitration procedure. The difference is in the arbitrator's solution. The arbitrator chooses

$$a_l \text{ if } a_l < 1 - a_m, \qquad (17)$$

and

$$a_m \text{ if } a_l > 1 - a_m$$

and his award. Therefore, he chooses that offer which demands a more modest payoff. (If the offers make equal demands, randomization decides which is chosen.)

The returns to players L and M are therefore

$$R_l^F(a_l, a_m; c_l) = \left(\frac{a_m + a_l}{2}\right)\delta(a_l, a_m)$$
$$+ (a_l - c_l)\gamma(a_l, a_m)(1 - \delta(a_l, a_m))$$
$$+ (a_m - c_l)(1 - \gamma(a_l, a_m))(1 - \delta(a_l, a_m)) \qquad (18)$$

and

$$R_m^F(a_l, a_m; c_m) = \left(1 - \left(\frac{a_l + a_m}{2}\right)\delta(a_l, a_m)\right)$$
$$+ (1 - a_l - c_m)\gamma(a_l, a_m)(1 - \delta(a_l, a_m))$$
$$+ (1 - a_m - c_m)(1 - \gamma(a_l, a_m))(1 - \delta(a_l, a_m)) \qquad (19)$$

where

$$\gamma(a_l, a_m) = 1 \text{ if } a_l < 1 - a_m,$$
$$= 0 \text{ if } a_\omega > 1 - a_m, \qquad (20)$$

and

$$\delta(a_l, a_m) \text{ is defined as before.}$$

In the next section we calculate how the players' strategies differ under the two modes of arbitration when costs are distributed uniformly. We expect player L's offer to decline as his cost rises and player M's offer to rise with his cost. The rate of that change in offer as cost increases could be interpreted as a concession rate (though dynamics are not explicitly built into the model), since the cost to a bargaining party goes up as the time spent in bargaining increases. If we assume that these costs will be borne at the end of the bargain (if there is no agreement), we obtain a model like ours.

B. Equilibrium Analysis of the Two Procedures

Proposition 5: One equilibrium pair of strategies under conventional arbitration is

$$a_l = x = a_m, \qquad \text{for all } c_l, c_m > 0. \qquad (21)$$

Proof: Suppose player M offers $a_m = x$. Player L would not then offer anything less than x, since he could always gain by moving up to x. If he offers anything more than x, $a_l > a_m$, and there is a disagreement so that his payoff reduces to $x - c_l \leqslant x$, for $c_l \geqslant 0$. The same reasoning holds for player M.

However, this equilibrium is not very interesting since it does not exploit a player's knowledge about his costs and his beliefs about the other player's costs. Intuitively one would expect such information to be used in the bargaining phase by players each seeking his individual advantage, with a low-cost player seeking to exploit his advantage by holding out for more.

In order to derive an equilibrium taking the differential cost aspect into *account*, we make the following assumptions.

1) Player L's strategy is a function $A_l(\cdot)$ which is strictly decreasing in c_l.

2) Players M's strategy is a function $A_m(\cdot)$ which is strictly increasing in c_m.

3) The probability distributions $F_m(\cdot), F_l(\cdot)$ have density functions everywhere.

4) The inverse functions $A_l^{-1}(\cdot)$ and $A_m^{-1}(\cdot)$ are differentiable everywhere.

Proposition 6: Given the assumptions about the model, an equilibrium pair of strictly monotone strategies under conventional arbitration will satisfy the following linked differential equations:

$$(x - c_l - a_l) = -\frac{1}{2} \frac{\left(1 - F_m(A_m^{-1}(a_l))\right)}{f_m(A_m^{-1}(a_l))A_m^{-1\prime}(a_l)}, \qquad (22)$$

and

$$(a_m - x - c_m) = \frac{1}{2} \frac{1 - F_l(A_l^{-1}(a_m))}{f_l(A_l^{-1}(a_m))A_l^{-1\prime}(a_m)}. \qquad (23)$$

Proof: Define Player L's conditional expected return given c_l and $F_m(\cdot)$ as

$$\bar{R}_l^c[a_l, A_m(\cdot) \mid F_m(\cdot), c_l] = \int_{A_m^{-1}(a_l)}^{\infty} \frac{a_l + A_m(c_m)}{2} f_m(c_m) dc_m$$
$$+ \int^{A_m^{-1}(a_l)} (x - c_l) f_m(c_m) dc_m. \quad (24)$$

In order to determine the maximum, we differentiate this and set it equal to zero, thus obtaining

$$\frac{1}{2}\left(1 - F_m(A_m^{-1}(a_l))\right) - a_l f_m(A_m^{-1}(a_l))A_m^{-1\prime}(a_l)$$
$$+ (x - c_l) f_m(A_m^{-1}(a_l))A_m^{-1\prime}(a_l) = 0. \quad (25)$$

Rearranging terms gives us (22). Equation (23) is derived analogously.

The quantity $(x - c_l)$ acts as a reservation price for player L, and it is partially unknown to player M, so that the framework set up here for the analysis of conventional arbitration is an extension of the distributive bargaining model analyzed elsewhere. However the extension enables us to obtain equilibria like that in Proposition 5 which is not possible in the earlier analysis.

From the point of view of conventional arbitration, the possibility of obtaining an equilibrium in monotone strategies may explain why agreements are not always reached in the bargaining phase. One would have been led to expect if Proposition 5 were the only equilibrium (other than the unappealing "no-bargain" kind) that the arbitration recourse would never be used. This is, of course, not empirically borne out.

In the next result we look at a special distribution of costs, the uniform distribution, which we shall use here for the purpose of comparing conventional arbitration and final-offer arbitration. First, we derive the equilibria under conventional arbitration.

Proposition 7: Let $F_m(\cdot)$ be uniform $[\alpha, \beta]$ and $F_l(\cdot)$ be uniform on $[\gamma, \delta]$. Then a pair of linear equilibrium interior

strategies are

$$a_l = A_l(c_l) = x + \frac{\beta}{4} - \frac{\delta}{12} - \frac{2}{3}c_l, \qquad (26)$$

and

$$a_m = A_m(c_m) = x - \frac{\delta}{4} + \frac{\beta}{12} + \frac{2}{3}c_m. \qquad (27)$$

Proof: The proof follows from (22) and (23) on substituting in the assumptions about $F_m(\cdot)$, $F_l(\cdot)$ and the linearity of the strategies.

The factor "x" which represents the arbitrator's intervention in the bargaining process can be manipulated by the arbitrator to shift the possible distribution of the total pie in favor of either labor or management. It cannot, however, affect the slope of the strategies (which we have interpreted, somewhat artificially, as a rate of concession in the bargaining). Since it occurs in exactly the same way in both players' strategies, it cannot be used to increase the probability of agreement.

Note that, as pointed out by a referee, the strategies given by (26) and (27) do not depend on the lower bounds α, γ of the two distributions. The mathematical reason for this is the form of (22) and (23) where the distributions enter in the form of $(1 - F)/f$, so that the lower bounds cancel where F is uniform. Intuitively, (26) and (27) say that each player is worried only about how high the other player's costs could go and about how high the other player thinks his or her costs could go. Thus player L would increase his demand if he knew that β was high and would decrease his demand if the upper bound of his own cost distribution, namely δ, were to increase.

We now look at final-offer arbitration, where the arbitrator selects one of the two final offers in accordance with the mechanism postulated in the previous section. Player L's expected return given c_l under FOA is written as

$$\bar{R}_l^F(a_l, A_m(\cdot) \mid F_m, c_l) = \int_{A_m^{-1}(a_l)}^{\infty} \frac{a_l + A_m(c_m)}{2} dF_m(c_m)$$
$$+ \int_{A_m^{-1}(1-a_l)}^{A_m^{-1}(a_l)} (A_m(c_m) - c_l) dF_m(c_m)$$
$$+ \int^{A_m^{-1}(1-a_l)} (a_l - c_l) dF_m(c_m). \quad (28)$$

The necessary conditions for optimality in this case do not have the same relatively neat expression as they do in the conventional arbitration case since they involve both $A_m^{-1}(a_l)$ and $A_m^{-1}(1 - a_l)$ in the arguments of both $F_m(\cdot)$ and f_m. For reference, they are given in Proposition 8.

Proposition 8: The necessary conditions for equilibrium strategies $A_l(c_l)$ and $A_m(c_m)$ under FOA are as follows:

$$\frac{1}{2}\left(1 - F_m(A_m^{-1}(a_l))\right) + F_m(A_m^{-1}(1 - a_l))$$
$$- c_l f_m(A_m^{-1}(a_l))A_m^{-1\prime}(a_l)$$
$$+ f_m(A_m^{-1}(1 - a_l))A_m^{-1\prime}(1 - a_l)(1 - 2a_l) = 0, \quad (29)$$

108   IEEE TRANSACTIONS ON SYSTEMS, MAN, AND CYBERNETICS, VOL. SMC-11, NO. 2, FEBRUARY 1981

and

$$-\frac{1}{2}\left(1-F_i\left(A_i^{-1}(a_m)\right)\right)-F_i\left(A^{-1}(1-a_m)\right)$$
$$-c_m f_i\left(A_i^{-1}(a_m)\right)A_i^{-1\prime}(a_m)$$
$$+(2a_m-1)f_i\left(A_i^{-1}(1-a_m)\right)A_i^{-1\prime}(1-a_m)=0. \quad (30)$$

Proof: These follow by differentiating the expressions for conditional expected returns and setting the result equal to zero.

Note that under this mode of arbitration there is no fixed focal point like "x" in our version of conventional arbitration. One would therefore expect the players to use monotone strategies, and if they were players well-versed in noncooperative game theory, the equilibrium monotone strategies given by (29) and (30). We now go to the uniform distribution special case with $F_m(\cdot)$ uniform on $[\alpha, \beta]$ and $F_i(\cdot)$ uniform on $[\gamma, \delta]$.

Proposition 9: An equilibrium pair in linear strategies with the above distributional assumptions is given by

$$a_i=A_i(c_i)=\frac{1}{12}\left(\frac{\beta}{2}-\alpha\right)-\frac{1}{84}\left(\gamma-\frac{\delta}{2}\right)+\frac{1}{2}-\frac{2}{7}c_i,$$
$$\quad (31)$$

and

$$a_m=A_m(c_m)=\frac{1}{12}\left(\gamma-\frac{\delta}{2}\right)-\frac{1}{84}\left(\frac{\beta}{2}-\alpha\right)+\frac{1}{2}+\frac{2}{7}c_i.$$
$$\quad (32)$$

Proof: As in analogous results here, we apply (29) and (30) assuming that $A_i(c_i)$ is a strictly decreasing linear function and $A_m(c_m)$ is a strictly increasing linear function. This gives us the slope $2/7$ of the linear strategies and two equations relating the two intercepts. Solving these gives us (31) and (32).

In order to simplify our comparison even further, let us consider the case where $F_m(\cdot)$ and $F_i(\cdot)$ are identical uniform distributions on $[0, \beta]$.[4] The equilibrium strategies are then as follows.

Under conventional arbitration we have

$$a_i=A_i(c_i)=x+\frac{\beta}{6}-\frac{2}{3}c_i, \quad (33)$$

and

$$a_m=A_m(c_m)=x-\frac{\beta}{6}+\frac{2}{3}c_m. \quad (34)$$

Under FOA we have

$$a_i=A_i(c_i)=\frac{\beta}{21}+\frac{1}{2}-\frac{2}{7}c_i, \quad (35)$$

and

$$a_m=A_m(c_m)=-\frac{\beta}{21}+\frac{1}{2}+\frac{2}{7}c_m. \quad (36)$$

---

[4] The linear strategies given hold as long as the probability of agreement is less than one. As soon as this probability becomes one (e.g., for values of $c_e \geqslant \beta/2$), the strategies become constant. This, of course, does not affect the probability of agreement as calculated in the text.

Equations (33) to (36) illustrate two interesting properties.

1) If the arbitrator's choice x is $1/2$, we see that under FOA, players are less extreme in their offers, differing only by $2\beta/21$ at most, while under conventional arbitration, they could differ by as much as $\beta/3$.

2) However, while bargainers tend to come down at a rate $2/3$ under conventional arbitration, they only do so at a $2/7$ rate under FOA. So flexibility in negotiation is not a virtue of FOA in this model, contrary to the claims of its sponsors.

If the probability of agreement is the criterion used to judge the two procedures, FOA comes out better than conventional arbitration. Under FOA there will be agreement if the sum of the costs exceeds $\beta/3$. In conventional arbitration agreement will take place when the sum of the costs exceeds $\beta/2$, a greater quantity.

These conclusions have been derived from considering a particular example. Different models of FOA and conventional arbitration and fewer heroic assumptions might yield different results. We intend to follow this up in future research.

V. CONCLUSION

We have discussed models of FOA and conventional arbitration under different informational assumptions. Our results would not cause us to recommend clear advocacy of one form of arbitration over another. However we feel that we have pointed out the inherent instability in final-offer arbitration, especially in Proposition 4. The model on which that proposition is based is particularly simple and surely reflects an essential feature of real-life bargaining—uncertainty about the arbiter's choice. The fact that this uncertainty is capable of causing extreme stands in bargaining is, or should be, disquieting. The distributional assumption (the uniform distribution) is certainly not unchallengeable, but neither does it appear implausible since bargainers might be expected to know the range of "fair" outcomes, but not which one the arbiter prefers.

Section IV of our paper, on models with different information, is an attempt to construct a model where the probabilities of agreement (in bargaining) under different arbitration procedures can be calculated explicitly. The author believes that there are very few other models in existence, if any, that enable us to do that, though as Crawford ([5]) indicates, research in this direction is urgently called for. Once again we assume the uniform distribution (once again surely not a wild assumption) to obtain explicit results. (Typically, bargaining models with differential information tend to have equilibria that are difficult to determine explicitly unless an analytically tractable distribution is assumed.) Section IV allows us to give only half a vote for FOA, since though the probability of agreement is higher under FOA, it appears to lead to a slow concession rate during the bargaining.

Our efforts in the direction of comparing the two procedures are, of course, still preliminary. A complete model

IEEE TRANSACTIONS ON SYSTEMS, MAN, AND CYBERNETICS, VOL. SMC-11, NO. 2, FEBRUARY 1981

would include risk-aversion as well as the trade-off between efficiency *after* disagreement and efficiency ex ante (before the process begins).

ACKNOWLEDGMENT

I wish to acknowledge the invaluable help of Howard Raiffa, who kindled my interest in this subject, and James Healy, who shared some of his experiences as an arbiter with me as well as introducing me to part of the industrial relations literature. I also wish to thank three anonymous referees for comments.

REFERENCES

[1] K. Chatterjee, "Interactive decision problems with differential information," unpublished DBA dissertation, Harvard Univ., Cambridge, MA, 1979.

[2] K. Chatterjee and W. Samuelson, "The simple economics of bargaining," mimeo, Pennsylvania State University and Boston University, 1979.

[3] K. Chatterjee and J. W. Ulvila, "Bargaining with shared information," mimeo, Pennsylvania State University and Decision Science Consortium, 1980.

[4] V. P. Crawford, "On compulsory arbitration schemes," *J. Political*

Econ., vol. 87, pp. 131–159, Feb. 1979.

[5] ——, "Arbitration and conflict resolution in labor management bargaining," in *Papers and Proc.*, American Economic Association Meeting, Sept. 1980, to appear.

[6] Henry S. Farber and H. C. Katz, "Interest arbitration, outcomes and the incentives to bargain," *Industrial, Labor Relations Rev.*, vol. 33, no. 1, Oct. 1979.

[7] P. Feuille, "Final-offer arbitration: concepts, developments, techniques," International Personnel Management Association, Chicago, IL, 1975.

[8] T. A. Kochan, R. G. Ehrenberg, J. Baderschneider, T. Jick, and M. Mironi, "An evaluation of impasse procedures for police and firefighters in New York State," Cornell Univ., New York State School of Industrial and Labor Relations, 1976.

[9] D. B. Lipsky, T. A. Barocci with W. Suoyanen, "The impact of final-offer arbitration in Massachusetts," mimeo, Massachusetts Institute of Technology, Cambridge, MA., 1977.

[10] R. Duncan Luce and Howard Raiffa, *Games and Decisions*. New York: Wiley, 1957.

[11] R. B. Myerson, "Incentive compatability and the bargaining problem," *Econometrica*, vol. 47, pp. 61–73, Jan. 1979.

[12] R. W. Rosenthal, "Arbitration under uncertainty," in *Rev. Economic Studies*, vol. 15, pp. 595–604, 1979.

[13] P. C. Somers, "An evaluation of final-offer arbitration in Massachusetts," Massachusetts League of Cities and Towns, Boston, MA., 1976.

[14] J. L. Stern, C. M. Rehmus, J. J. Loewenberg, H. Kasper, and B. D. Dennis, *Final-Offer Arbitration*. Cambridge, MA: Lexington Books, 1975.

INCENTIVE COMPATIBILITY IN BARGAINING UNDER UNCERTAINTY*

KALYAN CHATTERJEE

I. A MODEL OF BARGAINING UNDER UNCERTAINTY AND THE EFFICIENCY PROBLEM

The objective of this paper is to examine the question of designing efficient bargaining procedures in the context of a two-player model where each player has private information unavailable to the other.[1]

One simple instance of the kind of bargaining situation we wish to consider is provided in negotiations between a buyer and a seller who have to determine a price at which a single indivisible good is to be exchanged. Each participant typically has a reservation price that expresses his subjective valuation of the good (or embodies alternative opportunities available to that participant). This reservation price is not known to the other bargainer, though it is known to the player himself. We assume that these reservation prices are independent of income effects.

In the model, the true reservation prices are denoted by t_1 and t_2 for the buyer and seller, respectively. Using the now usual Harsanyi [1967–1968] formulation, we may think of t_1 and t_2 as random drawings from independent distributions $F_1(t_1)$ and $F_2(t_2)$ that are common knowledge in the sense of Aumann [1976].

Each player announces a value that he or she claims is his or her own reservation price. If the buyer's announced value (a_1) is greater

* I wish to acknowledge gratefully comments and suggestions from Takao Kobayashi, Keith Ord, John Pratt, Howard Raiffa, William Samuelson, Les Servi, two anonymous referees, and the editors of this *Journal*. Part of this research is included in my doctoral dissertation [1979] and was partially supported by the Defense Advanced Research Projects Agency, under contract N0014-76-0074 to Harvard University.

1. Our work differs from that of Myerson [1979] and Rosenthal [1978], which also address the efficiency problem, in that we seek to determine these procedures explicitly. The spirit (and content) of our results is closer to those of Chatterjee, Pratt, and Zeckhauser [1978]; Laffont and Maskin [1978]; Pratt and Zeckhauser [1979]; Samuelson [1979]; and the literature on the revelation problem for public goods described in Green and Laffont [1979]. Propositions 3 and 4 are applications to this context of a mechanism developed by D'Aspremont and Gerard-Varet [1979]. The basic model that we shall use here has been analyzed in detail in Chatterjee [1979], Chatterjee and Samuelson [1979], Chatterjee and Ulvila [1980], and Samuelson [1978]. The results of this paper have been generalized in a recent working paper by Myerson and Satterthwaite [1981].

© 1982 by the President and Fellows of Harvard College. Published by John Wiley & Sons, Inc.
The Quarterly Journal of Economics, November 1982   CCC 0033-5533/82/040717-10$02.00

than the seller's (a_2), a trade takes place at a price depending on a_1 and a_2. For example, it could be at $(a_1 + a_2)/2$, yielding payoffs of

(1) $t_1 - (a_1 + a_2)/2$ to Player 1

and

$(a_1 + a_2)/2 - t_2$ to Player 2.

If $a_1 < a_2$, there is no agreement, and both players get zero payoff.

Under the "split-the-difference" pricing procedure in (1), a profit-maximizing buyer will announce a value lower than his true reservation price in equilibrium, while a seller will announce one higher than his true value. Thus, on occasion, the buyer's announced reservation price might be lower than the seller's announcement, when, in fact, their true values were compatible. The players could therefore both end up with zero payoffs when they could both improve by being truthful.

In other words, if players announce truthfully, the sum of the payoffs is maximized for every pair of values of the reservation prices. This is, of course, as well as the players could have done with full information. Our aim here is to investigate the conditions under which truthful revelation is a Bayesian equilibrium strategy.

We proceed as follows: We first explore changes in the pricing procedure and find that there is no Bayesian incentive-compatible pricing procedure that satisfies a number of reasonable assumptions. An application of these negative results to bargaining on cost allocation is also discussed.

We then obtain a class of incentive-compatible schemes that do not, however, always ensure that the bargainers would wish to participate in the process if each bargainer knew his reservation price. A second incentive-compatible mechanism that suffers from a similar flaw is also introduced.

Note that in our framework (where a bargain takes place if and only if $a_1 \geq a_2$) the requirement of incentive compatibility is sufficient to guarantee full information efficiency.

II. INCENTIVE-COMPATIBLE PRICING PROCEDURES: IMPOSSIBILITY AND POSSIBILITY RESULTS

We define a pricing procedure $g(.\,,\,.)$ to be a function of the announcements a_1 and a_2 of the two players, respectively, that specifies the price at which the good is to be exchanged in the event of a bargain (i.e., when $a_1 \geq a_2$).

The payoff to the players on making announcements a_1 and a_2, respectively, is then given by $t_1 - g(a_1,a_2)$ to Player 1 and

$$(2) \qquad\qquad g(a_1a_2) - t_2 \qquad \text{to Player 2,}$$

if $a_1 \geq a_2$, and 0 otherwise.

In our statement of the payoff, we have already imposed one condition that will be relaxed later, namely, condition 1.

1. The Disagreement Payoffs Are Independent of the Players' Announcements

Thus, we shall be considering only "fixed-threat" type bargaining games in this section, where occurrence of a conflict will lead to fixed status quo payoffs.

Other intuitively plausible conditions that we could require our pricing procedure to satisfy are as follows:

2. Nonnegative Payoffs

Stated in words, this means that if the announcements are a_1 and a_2, respectively, and $a_1 \geq a_2$, then the transaction should take place at a price between the two announcements. Formally,

$$(3) \qquad\qquad a_1 \geq g(a_1,a_2) \geq a_2.$$

This ensures nonnegative profits for both players for all announcements $a_1 \leq t_1$ and $a_2 \geq t_2$. (As we have shown elsewhere, optimizing players will never announce values of a_1 and a_2 violating these restrictions.)

If the players agreed to this condition, they would have a positive incentive to play the game whatever their reservation prices. They would also never regret having participated in the bargain after the conclusion of the game. We could therefore also call this requirement that of *strong individual rationality*. This requirement is stronger than that of individual rationality which would require only positive conditional *expected* profit and would therefore be compatible with negative profit in some instances.

3. Unanimity

Notice that if $a_1 = a_2$, equation (3) implies that

$$(3') \qquad\qquad a_1 = g(a_1,a_1).$$

In other words, if the values of the two announcements coincide, then the transaction should take place at the common value. This is a slightly weaker assumption than the previous one but will prove sufficient for our results in this section.

4. Incentive Compatibility

That is, of course, the object of the whole exercise. We require that truth-telling strategies be in equilibrium; that is, the strategies

(4) $a_1 = t_1$ and $a_2 = t_2$

are best responses to each other.[2]

5. Positive Responsiveness to Announcements

This condition states that if a player's announced value becomes higher, the other player's announcement being fixed, the transaction price should also move upward. Formally,

(5) $\dfrac{\partial g}{\partial a_i} > 0,$ $\forall a_i,$ $i = 1,2.$

This requirement seems a plausible one, since it ensures that the procedure is responsive to the players' announcements for all values of a_1 and a_2.

6. Separability

This assumption states that the pricing procedure is separable in the announcements; i.e.,

(6) $g(a_1,a_2) = g_1(a_1) + g_2(a_2).$

These conditions may be considered desirable properties of a procedure, both individually and collectively.

In addition, we shall assume that $g(a_1,a_2)$ is at least twice continuously differentiable in both its arguments and that the distributions $F_1(\cdot)$ and $F_2(\cdot)$ are continuously differentiable everywhere in their range of definition.[3]

---

2. Note that this mutual best response property need hold only within certain relevant ranges, discussed in more detail in Chatterjee and Samuelson [1979]. Broadly speaking, these ranges are determined by the observation that an optimizing seller will never demand less than the lowest possible buyer offer and the irrelevance of changes in the announcement strategy that do not change the adversary's strategy or the probability of agreement.

3. Keith Ord has pointed out that conditions 1, 2, 6, and a symmetry assumption that $g_1 = g_2 = g_0$ imply the "split-the-difference" rule. The proof is as follows: From 6, $g(a_1,a_2) = g_1(a_1) + g_2(a_2)$. From 2, $g(a_1,a_1) = g_1(a_1) + g_2(a_1) = a_1$ and $g(a_2,a_2) = g_1(a_2) + g_2(a_2) = a_2$. From symmetry and the previous statement, $g_1 = g_2 = g_0$ and $2(g_0(a_1) + g_0(a_2)) = a_1 + a_2$, so that $g(a_1,a_2) = g_0(a_1) + g_0(a_2) = (a_1 + a_2)/2.$

PROPOSITION 1. Conditions 1, 2 (or 3), 4 and 5 are incompatible.

Proof. The expected return to the first player, given that the second tells the truth, is given by

$$(7) \qquad \overline{R}_1(a_1 | F_2, t_1) = \int_{-\infty}^{a_1} [t_1 - g(a_1, t_2)] f_2(t_2) dt_2.$$

The best a_1 can be found by differentiating and setting the resulting expression equal to zero, a procedure that gives us

$$(8) \qquad [t_1 - g(a_1, a_1)] f_2(a_1) = \int^{a_1} \frac{\partial g}{\partial a_1} f_2(t_2) dt_2.$$

By condition (3),

$$g(a_1, a_1) = a_1;$$

and by condition (4),

$$a_1 = t_1 \text{ is a best response.}$$

This gives us

$$(9) \qquad 0 = \int^{t_1} \frac{\partial g}{\partial a_1} f_2(t_2) dt_2.$$

Equation (9) violates condition 5. Hence, the result.

We can also show a related result that dispenses with condition 5, which is probably the most questionable of our conditions.

PROPOSITION 2. Conditions 1, 2 (or 3), 4, and 5 are incompatible.

Proof. Integrate equation (9) by parts. We then obtain

$$(10) \qquad 0 = \frac{\partial g}{\partial a_1}(a_1, a_1) F_2(a_1) - \int_{-\infty}^{a_1} \frac{\partial^2 g}{\partial t_2 \partial a_1} F_2(t_2) dt_2.$$

If the resolution procedure satisfies condition 6,

$$g(a_1, t_2) = g_a(a_1) + g_2(t_2),$$

so that

$$(11) \qquad \frac{\partial^2 g}{\partial t_2 \partial a_1} = 0.$$

Substituting (11) back into (10), we obtain

$$(12) \qquad 0 = g_1'(a_1) F_2(a_1), \qquad \forall a_1.$$

But $F_2(a_1)$ is equal to zero only if the value of a_1 is less than or equal to the lower bound of the distribution $F_2(\cdot)$; let this be denoted by $\underline{t_2}$. Therefore, for all $a_1 > \underline{t_2}$,

$$(13) \qquad\qquad g_1'(a_1) = 0.$$

Similarly, we find for all $a_2 < \overline{t_1}$ (the upper bound of the support of F_1),

$$(14) \qquad\qquad g_2'(a_2) = 0.$$

If we consider values in the open interval $(\underline{t_2}, \overline{t_1})$, assuming this to be nonempty,[4] we have, from (13) and (14),

$$(15) \qquad\qquad g_1'(x) + g_2'(x) = 0.$$

But, from differentiating both sides of equation (3') under the separability assumption,

$$(16) \qquad\qquad g_1'(x) + g_2'(x) = 1, \qquad \forall x.$$

Equations (15) and (16) cannot be satisfied simultaneously, hence the result.

These two results indicate that devising an incentive-compatible bargaining procedure is possible only if one gives up one or more of several desirable assumptions. Before we actually go on to do this, we pause for a moment to consider the application of these propositions to cost allocation for a common project.

The problem here is as follows. There are two divisions of a company that are considering the construction of a joint facility (perhaps a distribution center) cost c. Since the divisions are profit centers, they have to determine some method of allocating the cost, and an obvious suggestion might be to divide the cost in proportion to the benefits accruing to each division. That is, the proportion of the cost g_1 allocated to the first division could be

$$(17) \qquad\qquad g_1 = a_1/(a_1 + a_2),$$

where a_1 and a_2 are the announced benefits to divisions 1 and 2, respectively. The facility would be constructed if $a_1 + a_2 \geq c$.

Notice that this cost allocation scheme contains incentives for a division both to overstate and understate the true value of the benefit. If a division obtains a sizable benefit from the facility, it might want to overstate the value of the benefit in order to ensure construction of the facility. However, it then increases its share of the cost.

---

4. This is an innocuous assumption, since if it failed to hold, rational bargainers would find no advantage in playing the game.

One could hope that these opposing incentives would cancel out, but they do not. This is no accident. No such allocation scheme would work, by an application of Proposition 1, provided that we modify equation (3′) to equation (3″) which is

(3″) $$cg_1(a_1,c - a_1) = a_1,$$

and, of course,

$$g_1(a_1,a_2) + g_2(a_1,a_2) = 1,$$

for $a_1 + a_2 \geq c$.

Equation (3″) requires that if the two announcements just add up to the cost, the allocation should be the announcements themselves. The proof of the impossibility of such an allocation follows from Proposition 1.

Though incentive-compatible pricing procedures satisfying our conditions are, in general, impossible, we can obtain bargaining schemes that are incentive-compatible by relaxing the conditions that the bargaining satisfy condition 1; that is, of fixed conflict payoffs and that the pricing procedure satisfy condition 2 or 3. We shall work, in what follows, with separable pricing procedures for which we proved a negative result in Proposition 2, since we do not want to impose condition 5 at this stage. We also allow for zero-sum payments from one player to another quite apart from the price paid for the good that is being exchanged. (Propositions 3 and 4, which follow, are applications of the scheme developed by D'Aspremont and Gerard-Varet [1979] in the public goods context.)

Suppose that the price paid for the good if there is an agreement is given by

(18) $$\hat{g}(a_1,a_2) = \hat{g}_1(a_1) + \hat{f}_2(a_2),$$

where $\hat{g}$ satisfies conditions 2 and 6. Let the side payment made by Player 2 to Player 1 be $\phi(a_1,a_2)$, where

(19) $$\phi(a_1,a_2) = \phi_1(a_1) - \phi_2(a_2).$$

We can then show (see Chatterjee, Pratt, and Zeckhauser [1979]) that the following proposition holds.

PROPOSITION 3. There exists an incentive-compatible side payment scheme given by

(20) $$\phi_1(a_1) = \int^{a_1} [\hat{g}_1(a_1) + \hat{g}_2(t_2) - t_2]dF_1(t_2)$$

and

$$(21) \qquad \phi_2(a_2) = \int_{a_2} [t_1 - \hat{g}_1(t_1) - \hat{g}_2(a_2)] dF_1(t_1).$$

Note that $\phi_1(a_1)$ expresses Player 1's expectation of Player 2's payoff given that Player 2 announces truthfully and Player 1 announces a_1. A similar interpretation holds for $\phi_2(a_2)$.

We can obtain similar incentive-compatible schemes by varying the nature of the rules of the bargaining. For example, suppose that the bargaining takes place as follows: The buyer agrees to pay the seller an amount depending solely upon the announcements a_1 and a_2 and not on the conclusion or otherwise of an agreement. The good changes hands if a_1 is at least as large as a_2; otherwise the seller retains the good. It is relatively simple to show

PROPOSITION 4. An incentive-compatible payment from the buyer to the seller in the institutional arrangement described above, where the amount paid does not depend upon whether a bargain takes place or not, is given by

$$(22) \qquad g(a_1, a_2) = \int^{a_1} t_2 f_2(t_2) dt_2 + \int_{a_2} t_1 f_1(t_1) dt_1,$$

where g is now paid by Player 1 to Player 2. (Note that g does not satisfy condition 2, and the bargaining process does not satisfy condition 1.)

Proof. Player 1's expected return for this scheme of payments is

$$(23) \qquad \int^{a_1} (t_1 - t_2) f_2(t_2) dt_2 - E_{t_2} \int_{t_2} t_1 f_1(t_1) dt_1.$$

It is clear from the nature of the first term that Player 1 would choose $a_1 = t_1$ as a best response to $a_2 = t_2$. Similarly, Player 2's payoff is

$$(24) \qquad E_{t_1} \int^{t_1} t_2 f_2(t_2) dt_2 + \int_{a_2} (t_1 - t_2) f_1(t_1) dt_1,$$

which is again maximized by setting $a_2 = t_2$.

Both the incentive schemes in Propositions 3 and 4 can be made individually rational, ex ante (that is, prior to the players' learning their reservation prices) by suitably chosen lump sum side payments not depending on the announcements. However, one cannot guarantee

INCENTIVE COMPATIBILITY IN BARGAINING 725

that bargainers will wish to participate in games of this form *after* they learn their reservation prices. This is a serious deficiency, since a player *is* his reservation price (for bargaining purposes), and conceiving of a situation prior to his knowing himself seems somewhat hypothetical (like the "original position" of social contract theorists). An example is given below.

Suppose that both $F_1(\cdot)$ and $F_2(\cdot)$ are uniform distributions over $[0,1]$. Then Player 1's expected payoff (from expression (23)) is

$$(25) \qquad \int_0^{t_1} (t_1 - t_2)dt_2 - \int_0^1 \left(\int_{t_2}^1 t_1 dt_1 \right) dt_2$$

$$(26) \qquad \qquad \qquad = \tfrac{1}{2}t_1^2 - \frac{1}{2} \cdot \frac{t_1^2}{2} - \frac{1}{3}.$$

If t_1 is less than $\sqrt{2/3}$, Player 1 will not want to participate in the game.

However, prior to Player 1's knowing his reservation price, his expected profit is $\tfrac{1}{6} - \tfrac{1}{3} = -\tfrac{1}{6}$. Player 2's expected profit is $\tfrac{1}{3}$. By a suitable lump sum payment of $[\tfrac{1}{6} + \epsilon]$, therefore neither player would have an incentive to drop out of the bargaining.

III. Conclusion

We have examined the issue of efficient bargaining procedures in a game where players have private information. In general, it is not possible to devise such procedures, given a bargaining game with fixed conflict payoffs. By relaxing some conditions, however, we obtain two bargaining schemes that are efficient but that suffer from the defect that one (or both) of the bargainers might have an incentive not to participate once the private information is known. It appears impossible to resolve this defect in our framework.

Pennsylvania State University

References

Aumann, R. J., "Agreeing to Disagree," *Annals of Statistics*, IV (1976), 1236–39.
Chatterjee, Kalyan, "Interactive Decision Problems with Differential Information," DBA thesis, Harvard University, 1979.
——, John W. Pratt, and Richard J. Zeckhauser, "Paying the Expected Externality for a Price Quote Achieves Bargaining Efficiency," *Economics Letters*, I (1978), 311–13.
——, and William Samuelson, "The Simple Economics of Bargaining," mimeo, 1979.
——, and Jacob W. Ulvila, "Bargaining with Shared Information," mimeo, 1980.

D'Aspremont, Claude, and L. A. Gerard-Varet, "Incentives and Incomplete Information," *Journal of Public Economics*, XI (1979), 25–45.
Green, Jerry R., and Jean-Jacques Laffont, *Incentives in Public Decision Making* (Amsterdam: North-Holland Publishing Company, 1979).
Harsanyi, John C., "Games of Incomplete Information Played by Bayesian Players," *Management Science*, XIV (1967–1968), 158–82, 320–34, 486–502.
Laffont, Jean-Jacques, and Eric Maskin, "A Differential Approach to Expected Utility Maximizing Mechanisms," mimeo, Ecole Polytechnique, 1978.
Myerson, Roger B., "Incentive Compatibility and the Bargaining Problem," *Econometrica*, XLVII (Jan. 1979), 61–73.
——, and M. Satterthwaite, "Efficient Mechanisms for Bilateral Trading," Discussion Paper, Northwestern University, May 1981.
Pratt, John W., and Richard J. Zeckhauser, "Incentives Based Schemes for Decentralization," mimeo, 1979.
Rosenthal, Robert W., "Arbitration of Two-Party Disputes under Uncertainty," *Review of Economic Studies*, XLV (1978), 595–604.
Samuelson, W., "Models of Competitive Bidding under Uncertainty," Ph.D. thesis, Harvard University, 1978.
——, "An Efficient Solution to a Problem of Bilateral Monopoly under Uncertainty," mimeo, Boston University, 1979.

BARGAINING UNDER TWO-SIDED INCOMPLETE INFORMATION: THE UNRESTRICTED OFFERS CASE

KALYAN CHATTERJEE and LARRY SAMUELSON

Pennsylvania State University, University Park, Pennsylvania
(Received June 1986; revision received June 1987; accepted September 1987)

Chatterjee and Samuelson (1987) recently examined a noncooperative game-theoretic bargaining model with two-sided incomplete information and an infinite horizon. Results were obtained from the model with the help of restrictions on agents' strategies. This paper examines the model without these restrictions. By doing so, we will gain some insight into whether these results are robust, in the sense that they do not depend upon the special structure of the model, and hence, might provide useful theoretical foundations for applied work. We find that the basic results generalize. As with the restricted model, we find an equilibrium in which bargaining will proceed for a finite but endogenously determined number of stages. A terminal condition on the equilibrium sequence of agents' beliefs determines the properties of the equilibrium and allows comparative static results to be obtained. These include the finding that agents are more likely to capture a large share of the potential gains from bargaining if they exhibit characteristics that can be readily interpreted as giving them more bargaining power. Unlike the restricted model, this equilibrium is not unique, though we argue that it is plausible.

C hatterjee and Samuelson (1987) recently examined a noncooperative bargaining game with an infinite horizon and two-sided incomplete information. Strong results are derived with the help of a restrictive set of assumptions on bargainers' strategies. This paper examines the consequences of removing these restrictions.

Models of bargaining are most valuable if they allow progress to be made toward constructing theoretical foundations for applied work. Such progress requires the investigation of whether a model's results depend crucially upon special, perhaps arbitrary, characteristics of the model or whether the results are robust to alterations in such special assumptions. The latter finding would clearly instill more confidence in the model as a foundation for applied work. This paper provides an investigation of this type for one aspect of the bargaining model of Chatterjee and Samuelson (1987), the restrictive assumptions placed on bargainers' strategies, and finds the basic results to be relatively robust to removal of these restrictions.

Section 1 describes the model and presents a summary of the issues and results. Section 2 describes the relationship of the model and analysis to the literature. Section 3 constructs an equilibrium, and Section 4 establishes some key comparative static properties of the equilibrium. Section 5 discusses alternative equilibria, and the conclusions are outlined in Section 6.

1. The Model

We consider a bargaining game between a seller, who owns a single indivisible unit of an object, and a potential buyer. The seller's reservation price is $\underline{s}$ with probability π_s^1 (in which case we say that the seller is soft); and $\bar{s}$ with probability $1 - \pi_s^1$ (a hard seller). The buyer's reservation price is $\bar{b}$ with probability π_b^1 (a soft buyer); and $\underline{b}$ with probability $1 - \pi_b^1$ (a hard buyer). The prior probabilities π_s^1 and π_b^1 are given exogenously and are common knowledge. We have

$$\underline{s} < \underline{b} < \bar{s} < \bar{b}. \tag{1}$$

In period one, the seller makes an offer. The buyer either accepts or rejects this offer. In the former case, the game ends. In the latter, the game proceeds to period two, with payoffs now discounted by discount factors D_s and D_b. The buyer then makes an offer, which the seller either accepts (ending the game) or rejects. In the latter case, the sequence of offers begins again with payoffs now discounted by D_s^2 and D_b^2.[1]

In Chatterjee and Samuelson (1987), this game is examined under some restrictions, the most obvious of which is that offers come from the two-element set $\{\underline{b}, \bar{s}\}$. The following results appear. The game generically has a unique Nash equilibrium. Bargaining proceeds for a finite but endogenously determined number of periods. Hard sellers always offer $\bar{s}$; hard

Subject classification: Games/group decisions, noncooperative bargaining with incomplete information.

Operations Research
Vol. 36, No. 4, July–August 1988

0030-364X/88/3604-0605 $01.25
© 1988 Operations Research Society of America

buyers offer $\underline{b}$. With the possible exception of period one, soft agents randomize between the two offers in each period. As the game proceeds, the agents revise downward their expectation that their opponent is soft. The game must end by having the resulting sequence of probabilities hit a boundary in a certain way. Comparative static results can be derived.

While these results are interesting, the restriction on offers appears to play a crucial role in the results. The contribution of this paper is to show that the salient features of the equilibrium in the restricted offers game do not depend upon the offer restrictions and that they survive generalization of the model. We examine an equilibrium in the unrestricted offers game in which the basic features of the restricted offers equilibrium carry over. The bargaining proceeds for an endogenously determined, finite number of periods. Hard agents play pure strategies. Soft agents randomize in each period between an offer that reveals them to be soft and an offer that duplicates a hard agent, and hence, conceals their type. The equilibrium is determined by the necessity of having the revised probability-of-soft-opponent sequence hit a boundary in a certain way. Similar comparative static results can be derived.

Other equilibria also exist in the generalized game, supported by alternative beliefs at out-of-equilibrium events. The equilibrium we examine involves "optimistic" conjectures, which can be criticized. However, the alternative conjecture structures and equilibria we have investigated appear to have even less plausible features. These are discussed in Section 5 along with the issues involved in establishing uniqueness of equilibrium.

2. Relationship to the Literature

This paper constructs a noncooperative bargaining model involving two-sided incomplete information and an infinite horizon.[2] The results are closely related to three groups of papers. First, the play of the game may lead to a subgame of complete information. A complete-information bargaining game has been examined by Rubinstein (1982), and our solution to this subgame matches his.

Second, play may lead to a subgame of one-sided incomplete information. A similar game has been examined by Grossman and Perry (1986b). Grossman and Perry find multiple sequential equilibria, but isolate and examine a unique perfect sequential equilibrium (1986a). The latter equilibrium is similar to ours in that over time, the actions of the informed agent reveal information about the informed agent's type, with the uninformed agent revising beliefs accordingly. These beliefs become increasingly pessimistic and agents' offers become increasingly favorable to the informed agent, with the actions approaching the complete information outcome between the uninformed player and least favorable type of informed player.

The primary difference between the two equilibria is that in ours, the offers made by the informed agent are rejected along the equilibrium path. This behavior is supported by optimistic conjectures that deter the informed agent from making a more favorable offer. Grossman and Perry preclude optimistic conjectures, and some types of informed agents make offers which are accepted along the equilibrium path in their model. These differences appear to arise because Grossman and Perry allow a continuum of types of informed player while we allow only two such types. This seemingly technical difference becomes important when one investigates why out-of-equilibrium offers are not profitable for the *uninformed* agent. The presence of a continuum of informed agents in the Grossman-Perry model allows the specification of which *types* of informed agents accept to adjust to render out-of-equilibrium offers unprofitable. In our model with the two types, the *probability* with which the soft informed player accepts must adjust. This requires that soft agent indifference between acceptance and rejection be preserved, and this necessity imposes a constraint on the equilibrium which we satisfy through the use of optimistic conjectures.[3]

Third, the game begins with two-sided incomplete information. Analyses of similar games have been offered by Perry (1986) and Cramton (1986); the latter generalizes Cramton (1984). In Perry's model, bargaining proceeds for at most one stage. Either one agent quits the game immediately or one agent makes an offer which causes the other to either accept or quit. This result appears because the bargaining costs in Perry's model take the form of stage costs attached to each offer rather than discounting. Rubinstein (1982) shows that in a complete-information model with stage costs, the equilibrium allocates all of the gains from bargaining to the low-stage-cost agent. Perry (1986) extends this result to show that the weakest valuation high-stage-cost agent can receive no gains from trade in the incomplete information model. There can, accordingly, be no weakest high-stage-cost agent who continues bargaining after one stage, and the result that bargaining proceeds for at most one stage follows directly.

The model most similar to ours is investigated by Cramton (1986). Cramton's model differs in that it allows a continuum of types of buyers and sellers, and allows an offer to be made by either agent at any time (subject to a minimum response time constraint). Cramton's equilibrium strategies are thus technically quite different from ours, but are very similar in spirit. Agents conceal their types by making nonserious offers and reveal their types by choosing how long to delay before making a serious offer. Low (high) valuation sellers (buyers) delay a relatively short time. Delay causes expectations to be revised, with each agent's expectations becoming more pessimistic over time. The result is a separating equilibrium in which agents with relatively large potential gains from trade conclude agreements sooner than agents with smaller expected gains. This is a process similar to our equilibrium in which the probability that a soft agent takes a revealing action increases over the course of the game.

Cramton's analysis also resembles ours in that a multitude of sequential equilibria appear. The equilibrium examined by Cramton does not rely upon optimistic conjectures, however, and satisfies a collection of restrictions on conjectures and strategies that preclude optimistic conjectures. This possibility again appears to be driven by the presence of a continuum of types. This allows an equilibrium to be constructed that deters relatively hard agents from attempting to trade too soon by exploiting the fact that the continuum of agents can trade at a continuum of points in time, a result similar to Grossman and Perry. We cannot do this with the two-type specification, and resort to optimistic conjectures.

The collective results of these papers and Chatterjee and Samuelson (1987) allow some insights into the implications of various models of the bargaining process. The resemblance of our equilibrium to that of Chatterjee and Samuelson (1987) suggests that the offer restrictions imposed in the latter game have much the same effect as imposing optimistic conjectures in an unrestricted offers game. The analysis of Chatterjee and Samuelson (1987) also revealed that in the presence of offer restrictions, the basic properties of the equilibrium are unaffected by whether one presumes there to be a continuum of types of each agent or two types of each agent. The comparison of the equilibrium developed in this paper with Grossman and Perry (1986b) as well as Cramton (1986) suggests that in the absence of offer restrictions, the specifics of the equilibrium are more sensitive to the specification of the number of types of each agent.

3. Equilibrium

3.1. General Structure

This section provides an intuitive outline of the basic features of the equilibrium. We begin by identifying three possible phases of the game.

The game begins with a phase of two-sided incomplete information, with each agent unsure as to whether the opponent is hard or soft. The probabilities π_s^1 and π_b^1 give the prior expectations held by each agent concerning the opponent's identity. In the course of the play one agent, say the buyer, may take an action that reveals the agent's type (i.e., causes the seller to adopt a revised expectation given by $\pi_b = 0$ or $\pi_b = 1$). In this case, a second phase or subgame of one-sided incomplete information ensues between a buyer of known type and a seller of unknown type. In the course of this subgame, the seller may take an action that reveals the seller's type (yields a buyer expectation of $\pi_s = 0$ or $\pi_s = 1$), in which case a third phase or subgame of complete information between a buyer and seller of known type ensues.[4]

We now require some notation. Consider a complete information game. This game has a unique sequential equilibrium in which each agent is characterized by an offer the agent will make whenever allowed to do so. We let the following notation represent these offers:

Offer	Offering agent	Opponent
$\bar{\alpha}_s$	Hard seller ($\bar{s}$)	Soft buyer ($\bar{b}$)
$\underline{\beta}_b$	Soft buyer ($\underline{b}$)	Hard seller ($\bar{s}$)
$\bar{\alpha}_b$	Hard buyer ($\bar{b}$)	Soft seller ($\underline{s}$)
$\underline{\beta}_s$	Soft seller ($\underline{s}$)	Hard buyer ($\bar{b}$)
$\underline{\alpha}_b$	Soft buyer ($\underline{b}$)	Soft seller ($\underline{s}$)
$\underline{\alpha}_s$	Soft seller ($\underline{s}$)	Soft buyer ($\underline{b}$)

For example, $\bar{\alpha}_s$ is the optimal seller offer in a complete information game between hard seller ($\bar{s}$) and soft buyer ($\bar{b}$). To clarify this notation, notice that a subscript identifies the agent making an offer. An overbar indicates that the offering agent is hard; an underbar that the offering agent is soft. An offer against a soft opponent is indicated by an α; an offer against a hard opponent by β.

Rubinstein shows that $\bar{s} < \underline{\beta}_b < \bar{\alpha}_s < \bar{b}$ and $\underline{s} < \bar{\alpha}_b < \underline{\beta}_s < \underline{b}$. These reveal that offers lie between the valuations of the two agents and that the optimal buyer offer is lower than the optimal seller offer. We assume

$$\alpha_s < \bar{s} \tag{2}$$

$$\underline{b} < \underline{\alpha}_b \tag{3}$$

or graphically:

$$ \underset{\underline{s}}{\vdash}\quad\underset{\bar{\alpha}_b\ \beta_s\ \underline{b}}{+\!+\!+}\quad\underset{\alpha_b\ \alpha_s}{+\!+}\quad\underset{\tilde{s}}{+}\quad\underset{\beta_b\ \bar{\alpha}_s}{+\!+}\quad\underset{\bar{b}}{+} .$$

If $\underline{s}$ and $\tilde{s}$ are close to one another, (2) may fail. Similarly, (3) may fail if $\underline{b}$ and $\bar{b}$ are close. The assumption is thus a statement that the two types of buyer (seller) be sufficiently diverse. Its implication is that a hard agent will never want an opponent to believe that the hard agent is soft because the resulting complete-information-game payoff would be negative.

The basic features of the equilibrium can now be described. We begin with two-sided incomplete information. Soft agents would like to masquerade as hard ones, while hard agents would like to identify themselves. In each period, soft agents randomize between one offer that reveals them to be soft and one which conceals their type. The latter offer duplicates the offer made by the corresponding hard agent. The probabilities involved in the soft agent randomization in each period are chosen so as to support the previous-period soft agent randomization by making that agent indifferent between the outcomes of the two offers.

If a concealing offer appears, it is rejected. The appearance of such an offer, however, causes the opponent to revise downward the posterior probability describing the likelihood that the offerer is soft. A succession of concealing offers thus causes the posterior probabilities to march downward. This potentially continues until one of these probabilities crosses a boundary value. Once this boundary has been passed either a concealing offer is accepted or a revealing offer is made with probability one by the soft agent. Either event ends the game.

Before this boundary value is reached, one of the soft agent randomizations may yield a revealing offer. This initiates a subgame of one-sided incomplete information. In this subgame, the uninformed agent makes a series of offers. These offers become increasingly favorable to the informed agent, and are chosen to make the soft informed agent indifferent between accepting or rejecting so as to wait for the next, more favorable offer. The soft informed agent randomizes between accept and reject so as to support the optimality of the uninformed agent's offers. Both informed agents make concealing offers which are rejected by the uninformed agent. As this process continues, the uninformed agent revises downward the posterior probability describing the likelihood of a soft opponent (presuming that an offer acceptance

does not end the game first). This probability again passes a boundary value, at which point the game is immediately ended by the uninformed agent accepting the informed agent's offer.

3.2. Equilibrium Strategies

We now specify the equilibrium strategies precisely. We begin with hard agents.

Hard Buyer:

$$ \begin{cases} \text{Offer } \bar{\alpha}_b \\ \text{Accept offer } x \text{ iff } x \leqslant \beta_s \end{cases} \tag{4} $$

Hard Seller:

$$ \begin{cases} \text{Offer } \bar{\alpha}_s \\ \text{Accept offer } x \text{ iff } x \geqslant \beta_b. \end{cases} \tag{5} $$

These strategies duplicate the strategies followed by hard agents in a complete-information game against a soft opponent. We can thus think of hard agents as holding out for the payoffs they would receive from soft agents. We refer to $\bar{\alpha}_b$ ($\bar{\alpha}_s$) as a concealing offer for a buyer (seller).

We now consider soft agents, beginning with the soft buyer, and construct a sequence of offers:

$$ x_{0b} = \underline{s} + D_s(\bar{\alpha}_s - \underline{s}) \tag{6} $$

$$ x_{nb} = \underline{s} + D_s^2(x_{n-2b} - \underline{s}), \quad n = 2, 4, 6 \ldots . \tag{7} $$

The appearance of one of these offers will reveal the buyer to be soft, given (4)–(5), and we refer to these as revealing offers. These offers increase as n approaches 0, and the key characteristic of the offers is that a soft seller will be indifferent between accepting x_{nb} (for a payoff of $x_{nb} - \underline{s}$) or waiting two periods to accept the higher x_{n-2b} (for a payoff of $x_{n-2b} - \underline{s}$). Each of these offers will be the optimal offer for some value of π_s in the subgame of one-sided incomplete information between a soft buyer and a seller of unknown type. An analogous sequence of revealing offers for the soft seller exists:

$$ x_{0s} = \bar{b} + D_b(\bar{b} - \bar{\alpha}_b) \tag{8} $$

$$ x_{ns} = \bar{b} + D_b^2(\bar{b} - x_{n-2s}). \tag{9} $$

Recall that π_s^1 and π_b^1 are the original or prior probabilities of a soft opponent. In the course of the game, these expectations will be revised, and we let π_s^t and π_b^t be the period t probabilities. We then construct two increasing sequences of probabilities

$$ \{_{-1}\pi_b, {}_0\pi_b, {}_2\pi_b, {}_4\pi_b \ldots\} \tag{8a} $$

$$ \{_{-1}\pi_s, {}_0\pi_s, {}_2\pi_s, {}_4\pi_s \ldots\}. \tag{9b} $$

We will find that if an agent is to make a revealing offer in the two-sided incomplete information phase of the game, then the identity of that offer will depend upon where the posterior probability that the opponent is soft falls in sequence (8) or (9). The location of the probability in sequence (8) or (9) will also determine the offers made by the uninformed in a game of one-sided incomplete information. We next have sequences of probabilities denoted

$$\{h_s^1, h_s^3, h_s^5, \ldots\} \tag{10}$$

$$\{h_b^2, h_b^4, h_b^6, \ldots\}, \tag{11}$$

where $h_s^t (h_b^t)$ identifies the probability that a soft seller (buyer) will make a period t revealing offer in the two-sided incomplete information phase of the game. Finally, we have functions

$$k_s(x, \pi_s') \quad \eta_s(x) \tag{12}$$

$$k_b(x, \pi_b') \quad \eta_b(x). \tag{13}$$

The function $k_s(x, \pi_s')$ identifies, for any offer x made by the buyer in a subgame of one-sided incomplete information (with known, soft buyer), the probability that the soft seller accepts this offer. The function $\eta_s(x)$ will describe a randomization over offers made by the seller in certain out-of-equilibrium circumstances in a game of one-sided incomplete information with the seller known to be soft. The corresponding functions for the buyer are $k_b(x, \pi_b')$ and $\eta_b(x)$.

The Appendix proves the following proposition.

Proposition 1. There exist specifications for sequences (8)–(11) and the functions give in (12)–(13) such that the hard agent strategies given in (4)–(5) and the following soft buyer strategy and belief formation rule (the soft seller's strategy and beliefs are analogous) constitute a sequential equilibrium (cf., Kreps and Wilson 1982):

I. $\pi_b' < 1, \pi_s' < 1$ (Two-sided incomplete information):
   [Offers]. For even period t, offer $\bar{\alpha}_b$ with probability h_b' and offer x_{nb} (if $\pi_s' \in [_n\pi_s, _{n+2}\pi_s), n \geq 0$) or β_b (if $\pi_s' < _0\pi_s$) with probability $1 - h_b'$.
   [Accept/Reject]. For odd period t, reject any offer $x > \bar{\alpha}_s$. Accept $\bar{\alpha}_s$ if $\pi_s' \leq _{-1}\pi_s$.
II. $\pi_b' = 1$ (Known, soft buyer).
   [Offers]. For even period t, offer β_b if $\pi_s' < _0\pi_s$. Offer x_{nb} if $\pi_s' \in [_n\pi_s, _{n+2}\pi_s) (n \geq 0)$, unless the previous offer x in the one-sided incomplete information game came from the interval $[x_{n+2b}, x_{nb}) (n \geq 2)$, in which case offer x_{n-2b}

with probability $1 - \eta_b(x)$ and offer x_{nb} with probability $\eta_b(x)$.
   [Accept/Reject]. For odd period t, accept offer x iff $x \leq \alpha_s (x \leq \bar{\alpha}_s)$ whenever $\pi_s' \geq (<) 0^{\pi}s$, unless the previous buyer's offer in the one-sided incomplete information game came from the interval $[x_{2b}, x_{0b})$, in which case accept $x > (=) [<] \bar{\alpha}_s$ with probability $1 (\eta_b(x)) [0]$.
III. $\pi_s' = 1$ (Known, soft seller).
   [Offer]. For even period t, offer $\bar{\alpha}_b$.
   [Accept/Reject]. For odd period t, accept offer x with probability $k_b(x, \pi_b')$.
IV. Beliefs. Apply Bayes' rule whenever applicable. This applies to all situations except offers made by sellers. The buyer responds to any offer lower than the equilibrium offer by revising π_s' to one (if $\pi_s' > 0$), and hence, concluding the seller is soft.

3.3. Equilibrium Properties

Consider the initial, two-sided incomplete information phase of the game. As mentioned, hard agents essentially hold out for the payoff they would receive in a complete information game with a soft opponent. They do this by offering $\bar{\alpha}_s$ (hard seller) and $\bar{\alpha}_b$ (hard buyer). In light of this, soft agents have a choice between duplicating the hard agent offer or making an offer that reveals them to be soft. The latter offer will always be chosen so as to be the optimal initial offer in the one-sided game of incomplete information which it initiates. These offers are drawn from the sequences x_{nb} and x_{ns}, with x_{nb} (for example) being the optimal revealing offer in period t if $\pi_s' \in [_n\pi_s, _{n+2}\pi_s)$. The soft buyer, in general, randomizes between the concealing and revealing offer, with h_b' being the probability allocated to the concealing offer. A concealing buyer offer causes the seller to revise downward, via Bayes' rule, the probability that the buyer is soft. This phase of the game continues (unless a revealing offer appears) until either π_b' or π_s' is revised below $_{-1}\pi_b$ or $_{-1}\pi_s$. Consider the case in which π_b' falls below $_{-1}\pi_b$. At this point, the soft seller becomes so pessimistic concerning the possibility of extracting a favorable agreement from a soft buyer that the seller plays as if the buyer is known to be hard, either accepting the buyer's concealing offer or making the revealing offer of β_s (the optimal offer against a known, hard buyer) with probability one.

Before this can occur, the game may enter a one-sided incomplete information subgame with (for example) a known, soft buyer. We can presume that the buyer has the first move in this game, because the game will have been initiated by the buyer's making a revealing offer in the game of two-sided incomplete information.

The equilibrium proceeds as in Figure 1. In each period, the buyer makes an offer selected from the sequence x_{nb}. The offer selected depends upon the probability of a soft seller, with x_{nb} offered in period t if $\pi_s^t \in [_n\pi_s, _{n+2}\pi_s)$. These offers, which increase over time, are rejected by the hard seller. The soft seller randomizes between accepting and rejecting x_{nb}, with this randomization attaching a probability $k_s(x_{nb}, \pi_s^t)$ to accepting, which is calculated to cause π_s^t to be revised downward via Bayes' rule to $_{n-2}\pi_s$. The buyer's next offer will then be x_{n-2b}. Sellers make offers of $\bar{\alpha}_s$ which are rejected. This pattern continues until the buyer's expectation that the seller is soft reaches $_0\pi_s$. Here, the buyer offers x_{0b}, which the soft seller accepts with probability one. The hard seller rejects this in order to offer $\bar{\alpha}_s$ the next period, which is accepted.

One easily verifies the optimality of all actions along the equilibrium path of this subgame except the buyer's offers. These require some relatively intricate off-the-equilibrium path behavior. Suppose $\pi_s^t \in [_n\pi_s, _{n+2}\pi_s)$, so that x_{nb} is the prescribed equilibrium action. What if the buyer offers $x_{nb} - \epsilon$ for some small ϵ? To deter the buyer from making this offer, we might recommend that the soft seller reject $x_{nb} - \epsilon$. However, the attendant absence of probability revision will then induce the buyer to offer x_{nb} in period $t + 2$. The soft seller would rather accept $x_{nb} - \epsilon$ now (for sufficiently small ϵ), than wait two periods to receive an offer of x_{nb}, and the seller's rejection of $x_{nb} - \epsilon$ is thus sub-

optimal. To surmount this difficulty, the soft seller accepts offer $x_{nb} - \epsilon$ with probability $k_s(x_{nb} - \epsilon, \pi_s^t)$ ($< k_s(x_{nb}, \pi_s^t)$). This probability causes π_s to fall to $_n\pi_s$ (rather than $_{n-2}\pi_s$, as would occur with randomization $k_s(x_{nb}, \pi_s^t)$). The buyer, faced with $\pi_s^{t+2} = _n\pi_s$ then randomizes between x_{nb} (with probability η_{nb}) and x_{n-2b} (with probability $1 - \eta_{nb}$). Either offer is accepted with probability $k_s(x_., \pi_s^{t+2})$ and the game proceeds again along the equilibrium path. The probability η_{nb} is calculated to support the seller's random acceptance of $x_{nb} - \epsilon$. The sequence $_n\pi_s$ is calculated to ensure the buyer always prefers to make the equilibrium offer rather than a disequilibrium offer.[5]

4. Comparative Statics

We can now examine some properties of this equilibrium. We are especially interested in the relationship between agents' characteristics and the division of the expected gains from bargaining. We assume hereafter $\pi_s^1 > _{-1}\pi_s$ and $\pi_b^1 > _{-1}\pi_b$ in order to avoid continually having to note special cases.

Note that in the first two periods of the game, at least one of the soft buyer or seller must play a random strategy. The presumption that $\pi_s^1 > _{-1}\pi_s$ and $\pi_b^1 > _{-1}\pi_b$ ensures that neither agent will optimally allocate a probability of unity to a revealing offer. It similarly cannot be optimal for both agents to allocate a probability of unity to the concealing offer, since then the first two periods of the game will involve no possibility of trade and no probability revision. Both agents then bear discounting costs only to enter period three facing

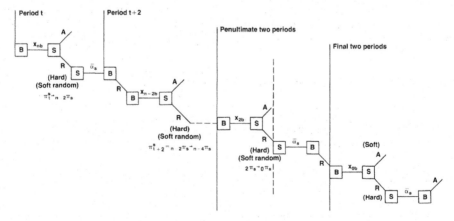

Figure 1. Equilibrium strategies.

a problem identical to that of period one. This cannot be an equilibrium, and at least one agent must then randomize in the first two periods. (See Lemmas A5–A7 in the Appendix for rigorous arguments.) Notice also that once this first random strategy has been played, each subsequent randomization in the two-sided incomplete information game is *uniquely* determined by the necessity of maintaining the indifference between revealing and concealing offers which supports the previous randomization.

In light of these observations, the equilibrium feature of key interest is whether the soft seller plays the first random strategy (in period one) or whether the soft seller plays a concealing offer with probability one in period one and the soft buyer plays the first random strategy (in period two). In particular, the agent who plays the first random strategy must be indifferent between playing a revealing and concealing offer, and hence, must receive a payoff equivalent to immediately revealing that the agent is soft. The agent who randomizes first does not exploit the incomplete information regarding its type to increase its payoff, while the other agent does. We can then say that the first agent to randomize does not receive any of the extra gains from trade made possible by the two-sided incomplete information game.

We would thus like to establish links between agents' characteristics and the identity of the first agent to randomize. Recall that the two-sided game ends when π'_s or π'_b falls below $_{-1}\pi_s$ or $_{-1}\pi_b$. Suppose that $\pi_s^{t-2} > {}_{-1}\pi_s$, but that the seller's period t offer randomization gives $\pi'_s < {}_{-1}\pi_s$. Two implications are then apparent. First, the soft buyer will accept $\bar{\alpha}_s$ in period t with probability one (since $\pi'_s < {}_{-1}\pi_s$). Secondly, the soft seller must have played a random strategy in period t in order to accomplish the probability revision from π_s^{t-2} to π'_s. In order for this random strategy on the part of the soft seller to be optimal, the payoffs from revealing or not revealing must be equal. Calculating these payoffs reveals that indifference requires $\pi'_b = {}_{-1}\pi_b$ (cf. Lemma A5 in the Appendix). Hence, the game can end with the soft buyer accepting $\bar{\alpha}_s$ only if the seller's probability revision that induces this acceptance occurs in a period in which $\pi'_b = {}_{-1}\pi_b$. Similarly, the game can end with the soft seller accepting $\bar{\alpha}_b$ only if the buyer's probability revision which induces this acceptance occurs in a period in which $\pi'_s = {}_{-1}\pi_s$. The first and second period randomizations of the agents (which determine all subsequent randomizations) must then be chosen so as to initiate a path of probabilities causing one agent's probability of being soft to drop below the boundary (i.e., π'_s must

fall below $_{-1}\pi_s$ or π'_b below $_{-1}\pi_b$) in a period in which the other agent's probability equals the boundary ($\pi'_b = {}_{-1}\pi_b$ or $\pi'_s = {}_{-1}\pi_s$). Figure 2 then describes possible equilibrium paths of the probabilities as well as paths which cannot yield equilibria.

Figure 2 suggests the following intuition concerning comparative statics. At point α, π_s^1 is relatively high and π_b^1 relatively low. It appears as if an equilibrium does not exist for this initial point in which the buyer randomizes first. Instead, the randomizations that characterize such a path decrease π'_b too rapidly relative to π'_s. The boundary of $_{-1}\pi_b$ is reached too soon (before π'_s is close enough to $_{-1}\pi_s$), precluding the existence of an equilibrium. What is needed here is for the seller to randomize first, and to undertake a randomization that significantly decreases π'_s. This allows π'_b to reach $_{-1}\pi_b$ when π'_s is sufficiently close to $_{-1}\pi_s$ to yield an equilibrium. Similarly, point β, with a high π_b^1 and low π_s^1, requires the buyer to randomize first. More formally, we have the following proposition.

Proposition 2. There exist increasing functions $\zeta_s(\pi_b)$ and $\zeta_b(\pi_s)$ such that if $\pi_s^1 < \zeta_s(\pi_b^1)$ ($\pi_b^1 < \zeta_b(\pi_s^1)$), then an equilibrium in which the seller (buyer) randomizes first generically does not exist.

Proof. Consider $\zeta_s(\pi_b^1)$. Generically (a qualification to be explained), an equilibrium path of probabilities initiated by a seller's randomization must arrive at interval I_1 in Figure 3. This interval contains all points such that the buyer's next randomization will yield $\pi'_b < {}_{-1}\pi_b$ (given $\pi_s^{t-1} = {}_{-1}\pi_s$ and given that the buyer's randomization is chosen to support the seller's previous period randomization). We can now identify interval or collection of intervals, I_2, which is the set of points from which a single buyer and then a seller's randomization (each of which is chosen to support the opponent's indifference between a concealing and revealing offer in the previous period) yield points on I_1. Continuing in this manner gives intervals I_1, I_2, ... For an equilibrium to exist in which the seller randomizes first, it must be the case that the seller's first randomization can convert π_s^1 into a probability π_s^2 such that the point (π_s^2, π_b^1) is contained in one of the intervals $I_1, \ldots$. Let $\zeta_s(\pi_b)$ be an increasing function that lies below all I_l. If $\pi_s^1 < \zeta_s(\pi_b)$, then π_s^1 would have to be increased by the seller's first randomization in order for (π_s^2, π_b^1) to be contained in an interval I_l, and hence, allow an equilibrium. However, no seller's randomization can accomplish this increase, and an equilibrium in which the seller randomizes first,

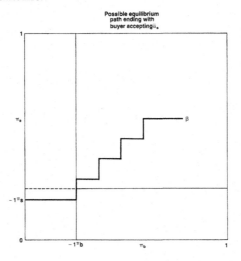

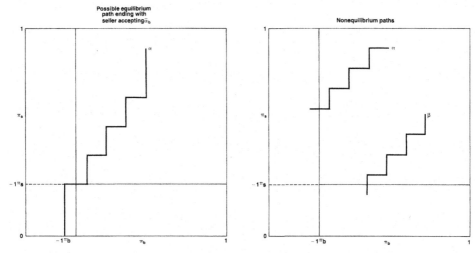

Figure 2. Equilibrium belief paths.

accordingly, does not exist. We similarly calculate $\zeta_b(\pi_s)$.[6]

This provides the desired comparative static result. There exists a set of initial probabilities, in which the probability of a soft buyer is high relative to that of a soft seller, for which the buyer must randomize first in equilibrium (hence, sacrificing the extra gains from trade to be had from engaging in incomplete-

information bargaining). There also exists a set of initial probabilities, in which the probability of a soft seller is high relative to that of a soft buyer, for which the seller must randomize first in equilibrium.

5. Conjectures and Uniqueness

We constructed and examined a particular equilibrium. This equilibrium is supported by optimistic

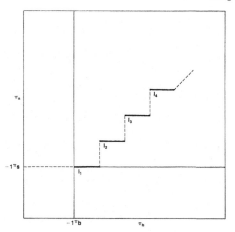

Figure 3. Construction of boundary on equilibrium types.

conjectures, which have been criticized. Grossman and Perry (1986a), for example, cite the general preclusion of such conjectures as an advantage of their perfect sequential equilibrium concept. (Our equilibrium is not a perfect sequential equilibrium, though it does satisfy the intuitive criterion of Cho and Kreps (1987).) The question then arises as to whether other equilibria also exist in our model.

We can first demonstrate that other equilibria exist that also feature the optimistic conjectures described above. To do this, consider the subgame of one-sided incomplete information with a soft buyer. Sever the equilibrium path shown in Figure 1 at the broken vertical line and attach Figure 4. We could then construct an equilibrium similar to that of Proposition 1. In each period, the sellers offer $\bar{\alpha}_s$, and the buyer rejects. The buyer makes offers which make the soft seller indifferent between each offer and the buyer's offer two periods hence. The soft seller randomizes in accepting or rejecting these offers so as to support the optimality of the buyer's offers. The buyer's sequence of offers approaches β_b rather than x_{0b}, so that the equilibrium path ends as in Figure 4 rather than

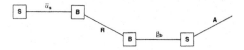

Figure 4. Alternative equilibrium path terminus.

Figure 1. This equilibrium is qualitatively no different than the one constructed above. However, we prefer the first equilibrium. Consider the two periods shown in Figure 4. By the time the game has proceeded this far, it is essentially a complete information game. The only actions that remain along the equilibrium path are actions which would be optimal in the complete information game. In addition, the probability that the seller is soft has fallen to such a low level that the buyer makes no attempt to take advantage of this possibility along the equilibrium path. In spite of being essentially a complete information game, however, the equilibrium calls for two offers to be made before an agreement is reached. The rejection of the first offer is supported by optimistic conjectures. It seems somewhat unreasonable that a conjecture should play such an important role in the equilibrium at this point, and this equilibrium accordingly appears to be problematic.[7] It is important to observe, however, that the presence of this equilibrium alters none of the results reported above.

If we relax the imposition of optimistic conjectures, then another equilibrium arises. Remove the final two

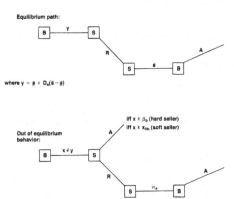

Figure 5. Alternative equilibrium path terminus without optimistic conjectures.

periods from the equilibrium path in Figure 1 and attach the equilibrium path shown in Figure 5. The calculations that determine $\{x_{nb}\}$ and $\{_n\pi_s\}$ are now performed with x_{0b} replaced by y, so that the buyer's sequence of equilibrium offers approaches y. The offer y is calculated so that the soft seller is indifferent between accepting y in (say) period t or having $\bar{s}$ accepted in period $t + 1$. It is easily verified that the hard seller prefers to have $\tilde{s}$ accepted in period $t + 1$. The sellers' period t equilibrium actions are thus optimal. In period $t + 1$, we presume that the buyer forms the following beliefs. If the buyer's period t offer was y, then any period $t + 1$ offer above $\tilde{s}$ is taken to indicate that the seller is soft with probability one. If the buyer's period t offer was not y, then any offer above $\tilde{\alpha}_s$ is taken to indicate that the seller is soft with probability one. Given these conjectures, the seller's period $t + 1$ offers are optimal (given that they are accepted). Any play beyond this period is assumed to occur according to the strategies given in Proposition 1, so that the buyer's acceptance is optimal.

It remains to verify that offer y is optimal in period t. By offering y, the buyer secures a relatively favorable offer of $\tilde{s}$, but must wait one period to do so. The alternative is to offer either x_{0b} or β_b in order to secure agreement more quickly at a higher offer. If the discount factor is high enough, so that the buyer is sufficiently patient, offer y will be optimal. We then have a third equilibrium.

The conjectures that support this third equilibrium exhibit two counterintuitive features. First, a higher seller offer is taken to indicate the presence of a soft seller, while intuition associates low offers with soft sellers. Secondly, the period $t + 1$ conjecture depends critically upon the buyer's period t offer, including various offers which yield equivalent period two situations. In effect, this conjecture structure allows the buyer to play a commitment strategy. It appears unreasonable that conjectures should allow the buyer to effectively play commitment strategies, and this equilibrium again appears problematic.

The multiplicity of equilibria suggests that restrictions be placed on conjectures, perhaps justified by appeal to an equilibrium refinement, in order to select a unique equilibrium. We have been able to accomplish this only by first imposing enough restrictions on conjectures to allow only optimistic conjectures and then imposing a "continuity" requirement, ensuring that conjectures become unimportant as beliefs approach certainty, which excludes the equilibrium identified in Figure 4. We find the resulting equilibrium appealing and believe that it satisfies a number of reasonable properties. However, the strength of the assumptions on conjectures makes this a weak uniqueness result. In light of this, we take the contribution of this paper to be the demonstration that the model has multiple equilibria, one of which appears to be plausible and which mimics the equilibrium of the restricted-offers game.

6. Conclusion

We generalized the model of Chatterjee and Samuelson (1987) by removing restrictions on agents' offers. The unique equilibrium of the restricted-offers game has been replaced by a multitude of equilibria. However, an equilibrium exists which shares the features of the restricted-offers game equilibrium. Bargaining proceeds for at most an endogenously determined, finite number of stages. Hard agents play pure strategies, and soft agents randomize in each period (with the possible exception of the first) between offers which duplicate those of hard agents and offers which reveal soft agents to be soft. The specifics of the equilibrium and the division of the gains from bargaining are determined by the necessity of having the sequence of updated beliefs hit a boundary in a certain way. The general features of the model's implications for bargaining also remain. We have established comparative statics results that link the game's parameters to the division of the gains from bargaining in a way which matches intuitive notions of bargaining strength.

Appendix

We prove Proposition 1 and develop results used in Section 5. We first demonstrate the optimality of the strategies in a subgame of one-sided incomplete information. Consider the case of $\pi_b = 1$, or a known, soft buyer. (The soft seller is analogous.) For convenience, let the initial probability that the seller is soft be denoted π_s^1.

Lemma A1. *Let*

$$
_0\pi_s \equiv \frac{(\bar{b} - \beta_b) - D_b(\bar{b} - \tilde{\alpha}_s)}{(\bar{b} - x_{0b}) - D_b(\bar{b} - \tilde{\alpha}_s)}
$$
$$
= \frac{(\bar{b} - \tilde{\alpha}_s)(1 - D_b^2)}{D_b(\bar{b} - x_{0b}) - D_b^2(\bar{b} - \tilde{\alpha}_s)}, \tag{A1}
$$

where $\beta_b = \tilde{s} + D_s(\tilde{\alpha}_s - \tilde{s})$ and $D_b(\bar{b} - \beta_b) = (\bar{b} - \tilde{\alpha}_s)$ by construction (cf. Rubinstein 1982) and x_{0b} is given by (6). Then, if $\pi_s^1 \in [0, {_0\pi_s})$, a sequential equilibrium (termed equilibrium A) is given by (5) and the following soft agent strategies (note $\beta_b > x_{0b}$, so the buyer makes a single offer of β_b which is accepted).

Buyer:

$\begin{cases} \text{Offer } \beta_b \text{ in odd periods.} \\ \text{Accept } x \text{ iff } x \leq \tilde{\alpha}_s \text{ in even periods.} \end{cases}$

Soft Seller:

$$\begin{cases} Accept\ x\ iff\ x \geq x_{0b}\ in\ odd\ periods. \\ Offer\ \bar{\alpha}_s\ in\ even\ periods. \end{cases}$$

Proof. By construction, $\beta_b\ (x_{0b})$ makes the hard (soft) seller indifferent between $\beta_b\ (x_{0b})$ in the present period and $\bar{\alpha}_s$ in the next. This suffices to yield the optimality of the sellers' strategies. The buyer has no incentive to make a higher period 1 offer, since the current one is accepted. Any lower offer precludes hard seller acceptance, but induces the soft seller to accept as long as $x \geq x_{0b}$. Any offer from $[x_{0b}, \beta_b)$ then has acceptance probability (π_s^1), and the best possible offer from this interval is thus the lowest, x_{0b}. Calculation shows that offer β_b dominates x_{0b} iff $\pi_s^1 \leq {}_0\pi_s$.

Lemma A2. *Let* $\pi_s^1 \in [{}_0\pi_s, {}_2\pi_s)$, *where* ${}_2\pi_s$ *solves*

$$(\bar{b} - x_{2b})_2\pi_s k(x_{2b}, {}_2\pi_s)$$

$$+ (1 - {}_2\pi_s k(x_{2b}, {}_2\pi_s))D_b(\bar{b} - \bar{\alpha}_s)$$

$$= (\bar{b} - x_{0b})_2\pi_s + (1 - {}_2\pi_s)D_b(\bar{b} - \bar{\alpha}_s), \qquad (A2)$$

with x_{2b} *defined by (7) and with* $k(x, \pi_s) = 1$ *if* $x \geq x_{0b}; = 0$ *if* $x < x_{2b};$ *and for* $x \in [x_{2b}, x_{0b})$:

$$k(x, \pi_s) = \frac{{}_0\pi_s - \pi_s}{{}_0\pi_s\pi_s - \pi_s}. \qquad (A3)$$

Then an equilibrium (termed equilibrium B) is given by (5) and the following soft agent strategies, coupled with the conjecture described in Proposition 1:

Buyer:

Offer x_{0b} *in odd periods.*

Accept $\bar{\alpha}_s$ *in even periods if* x_{0b} *was the previous offer. Reject offer* $x > \bar{\alpha}_s$.

Accept $\bar{\alpha}_s$ *in even periods with a probability* $\eta_b(x)$ *sufficient to make the soft seller indifferent between accepting and rejecting the buyer's previous offer if that offer comes from* $[x_{2b}, x_{0b})$.

Soft Seller:

Accept x *with probability* $k(x, \pi_s^1)$ *in odd periods.*

Offer $\bar{\alpha}_s$ *in even periods.*

Proof. It is clear that if an initial offer of x_{0b} is made, the ensuing behavior is optimal (noting that a seller is deterred from making an offer below $\bar{\alpha}_s$ by the conjecture described in Proposition 1). We demonstrate the optimality of the initial offer of x_{0b}. From Lemma A1, we know $x > x_{0b}$ is not optimal. What about $x < x_{0b}$? If an offer from $[x_{2b}, x_{0b})$ occurs, the soft seller randomizes between accepting and rejecting, with the acceptance probability $k(x, \pi_s^1)$ chosen to

lower π_s^1 to ${}_0\pi_s$. This makes the buyer indifferent between accepting and rejecting $\bar{\alpha}_s$ next period. The buyer randomizes between accepting and rejecting $\bar{\alpha}_s$ next period in order to make the soft seller indifferent between accepting and rejecting the buyer's offer in period one. To ensure that this is an equilibrium, we must ensure that the buyer does not prefer any offer from $[x_{2b}, x_{0b})$ to x_{0b}. The highest payoff offer from this interval is x_{2b}. Calculation of payoffs shows that offer x_{2b} will not be preferred to x_{0b} if $\pi_s^1 < {}_2\pi_s$ (cf. (A2) and (A3)). The soft seller will obviously reject an offer $x < x_{2b}$ since waiting two periods to accept x_{0b} is preferred.

Lemma A3. *Let a sequence* $\{x_0,\ x_2,\ x_4,\ x_6\ ...\}$ *be defined by (6)–(7). Let* $\{{}_0\pi_s,\ {}_2\pi_s,\ {}_4\pi_s,\ {}_6\pi_s,\ ...\}$ *and* $\{V_0,\ V_2,\ V_4,\ V_6,\ ...\}$ *be sequences defined by* ${}_0\pi_s$ *(cf. (A1)),* ${}_2\pi_s$ *(cf. (A2)–(A3)) and*

$$V_0 = (\bar{b} - x_{0b})_0\pi_s + D_b(1 - {}_0\pi_s)(\bar{b} - \bar{\alpha}_s) \qquad (A4)$$

$$V_n = (\bar{b} - x_{nb})_n\pi_s k(x_{nb},\ {}_n\pi_s)$$

$$+ (1 - {}_n\pi_s k(x_{nb},\ {}_n\pi_s))D_b^2 V_{n-2}, \qquad (A5)$$

$$(\bar{b} - x_{nb})_{n+2}\pi_s k(x_{nb},\ {}_{n+2}\pi_s)$$

$$+ (1 - {}_{n+2}\pi_s k(x_{nb},\ {}_{n+2}\pi_s))D_b^2 V_{n-2}$$

$$= (\bar{b} - x_{n+2b})_{n+2}\pi_s k(x_{n+2b},\ {}_{n+2}\pi_s) \qquad (A6)$$

$$+ (1 - {}_{n+2}\pi_s k(x_{n+2b},\ {}_{n+2}\pi_s))D_b^2 V_n,$$

where, for $\pi_s \in [{}_n\pi_s,\ {}_{n+2}\pi_s)$:

$$k(x, \pi_s)$$

$$= \begin{cases} 1 & if\ x \geq \beta_b \\ \dfrac{m\pi_s - \pi_s}{m\pi_s\pi_s - \pi_s} & if\ x \in [x_{m+2b},\ x_{mb})\ m < n \\ \dfrac{n-2\pi_s - \pi_s}{n-2\pi_s\pi_s - \pi_s} & if\ x = x_{nb} \\ \dfrac{n\pi_s - \pi_s}{n\pi_s\pi_s - \pi_s} & if\ x \in [x_{n+2b},\ x_{nb}) \\ 0 & if\ x < x_{n+2b} \end{cases} \qquad (A7)$$

Let $\pi_s^1 \in [{}_n\pi_s,\ {}_{n+2}\pi_s)$ *for* $n > 0$ *(strategies for* $\pi_s^1 \in [{}_0\pi_s,\ {}_2\pi_s)$ *are as in Lemma A2 and strategies for* $\pi_s^1 \in [0,\ {}_0\pi_s)$ *are as in Lemma A1; so that equilibria A and B are the special cases of this equilibrium which appear when* $\pi_s^1 < {}_2\pi_s$*). Then a sequential equilibrium is given by the following strategies and (5).*

Buyer:

Offer x_{nb} *in odd period t if* x_{n+2b} *was offered two periods ago.*

Randomize between offering x_n and x_{n+2} in period t
if $x \in [x_{n+4b}, x_{n+2b})$ was offered two periods ago,
with randomization $\eta(x)$ chosen to make soft seller
indifferent between accepting and rejecting x.
Accept x in even periods iff $x \leqslant \alpha_s$.

Soft Seller.

Offer $\bar{\alpha}_s$ in even periods.
Accept x_{nb} with probability $k(x_{nb}, \pi'_s)$ in odd period
t. This causes π'_s to fall to $_{n-2}\pi_s$.
Reject x if $x < x_{n+2b}$ in period t.
Accept x in period t if $x \in [x_{n+2b}, x_{nb})$ with proba-
bility $k(x, \pi'_s)$. This cause π'_s to fall to $_n\pi_s$.
Accept $x \geqslant x_{nb}$ with probability $k(x, \pi'_s)$, which is
sufficient to drop π'_s to $_m\pi_s$ if $x \in [x_{m+2}, x_m)$ for
$m < n$.

Proof. It is clear that the seller's strategies are optimal
given the buyer's strategy. It remains to show that x_{nb}
is preferred by the buyer to any other possible offer.
It is clear that x_{nb} dominates any offer $x < x_{n+2b}$.
Consider an offer in the interval $[x_{n+2b}, x_{nb})$. If offer
x_{nb} is made, the equilibrium calls for the seller to
accept with probability $k(x_{nb}, \pi'_s)$, reducing π'_s to $_{n-2}\pi_s$
and inducing offer x_{n-2b} two periods hence. If instead
$x \in [x_{n+2b}, x_{nb})$ is offered, the equilibrium calls for the
seller to accept with probability $k(x, \pi'_s)$, reducing π'_s
to $_n\pi_s$. At this probability, the buyer is indifferent
between offering x_{nb} and x_{n-2b} two periods hence. The
buyer can then randomize between x_{nb} and x_{n-2b} to
render the soft seller indifferent between accepting
and rejecting x, and hence, making the soft seller's
randomization optimal. Whichever offer appears in
the buyer's randomization, the game proceeds along
the appropriate equilibrium path. This yields an equilibrium as long as $x \in [x_{n+2b}, x_{nb})$, accepted with
probability $k(x, \pi'_s)$, is not preferred by the buyer to
offer x_{nb} accepted with probability $k(x_{nb}, \pi'_s)$. Conditions (A4)–(A7) ensure that the interval $[_n\pi_s, {}_{n+2}\pi_s)$,
for which offer x_{nb} is made, is such that no offer $x \in$
$[x_{n+2b}, x_{nb})$ will dominate offer x_{nb} (cf. (A6)). It is this
constraint that determines the sequence $\{_n\pi_s\}$. Offer
x_{nb} is then preferred to any $x < x_{nb}$. Similarly, if an
offer $x \in [x_{mb}, x_{m-2b})$ is made for $m < n$ $(x > x_{nb})$, the
seller randomizes to drop π'_s to $_{m-2}\pi_s$, with this randomization supported by a buyer randomization
between x_{m-2b} and x_{m-4b} at next opportunity.
The sequence $_n\pi_s$ again ensures that offer x_{nb} is not
dominated.

An analogous equilibrium arises in the one-sided
incomplete information game with a known, soft
seller. An implication is that there exists a step function $x_b(\pi_s)$ which identifies the optimal initial offer of

a soft buyer if a one-sided incomplete information
game is to be entered with a probability of a soft seller
of π_s. An analogous function $x_s(\pi_b)$ exists for the soft
seller. The existence of these functions is easily
exploited to prove the following lemma.

Lemma A4. *There exist continuous functions* $\phi_s(\pi_b) \cdot$
*($\phi_b(\pi_s)$), which give the expected payoff to a known,
soft seller (buyer) from entering a one-sided incomplete
information game with probability π_b (π_s) of a soft
opponent.*

We now consider the case of the two-sided incom-
plete information. The hard seller offers $\bar{\alpha}_s$ in each
period, and the hard buyer offers $\bar{\alpha}_b$ with these
offers supported by the conjecture described in Proposition 1.

The offers that soft agents might adopt are also
easily determined. If they wish to conceal their type,
the offer must be $\bar{\alpha}_s$ for a soft seller and $\bar{\alpha}_b$ for a soft
buyer, since these match the offers of hard agents. If
a revealing offer is to be made, the agents enter a game
of one-sided incomplete information. The revealing
offer must then be the optimal initial offer for such a
game, and will be $x_b(\pi'_s)$ for a buyer or $x_s(\pi_b)$ for a
seller. Soft sellers thus offer $\bar{\alpha}_s$, $x'_s(\pi_b)$, or a randomization between the two; soft buyers offer $\bar{\alpha}_b$, $x_b(\pi'_s)$,
or a randomization. Recalling that $x_b(\pi'_s)$ and $x_s(\pi'_b)$
are step functions whose image sets contain only a
finite number of values, we can then restrict attention
to a finite set of offers.

Lemma A5. *The hard seller (buyer) always rejects*
$\bar{\alpha}_b(\bar{\alpha}_s)$. *The soft seller (buyer) accepts* $\bar{\alpha}_b$ ($\bar{\alpha}_s$) *in period*
t *iff*

$$\pi'_b \leqslant \frac{(\bar{\alpha}_b - \underline{s})(1 - D_s^2)}{D_s(\bar{\alpha}_s - \underline{s}) - D_s^2(\bar{\alpha}_b - \underline{s})} \equiv {}_{-1}\pi_b < {}_0\pi_b \qquad (20)$$

$$\pi'_s \leqslant \frac{(\bar{b} - \bar{\alpha}_s)(1 - D_s^2)}{D_b(\bar{b} - \bar{\alpha}_b) - D_b^2(\bar{b} - \bar{\alpha}_s)} \equiv {}_{-1}\pi_s < {}_0\pi_s. \qquad (21)$$

Proof. Consider the buyer. Since $\underline{b} < \bar{\alpha}_s$, it is clear
that the hard buyer will reject $\bar{\alpha}_s$. Consider the soft
buyer. If the buyer accepts, the buyer receives a payoff
of $\bar{b} - \bar{\alpha}_s$. Rejecting $\bar{\alpha}_s$ reveals the buyer to be hard,
since only hard buyers remain in the game (in equilib-
rium) in period $t + 1$. The soft seller then accepts $\bar{\alpha}_b$
(the equilibrium offer in a complete information game
between soft seller and hard buyer) and the soft buyer's
payoff is $D_b\pi'_s(\bar{b} - \bar{\alpha}_b) + D_b^2(1 - \pi'_s)(\bar{b} - \bar{\alpha}_s)$. Rearranging reveals that this will be dominated by $\bar{b} - \bar{\alpha}_s$
iff $\pi'_s \leqslant {}_{-1}\pi_s$. The analysis of the seller's decisions and
$_{-1}\pi_b$ is similar.

Lemma A6. *If $t > 1$ and the period t probability that the seller (buyer) is soft exceeds $_{-1}\pi_s$ $(_{-1}\pi_b)$, then the soft seller (buyer) cannot play $\tilde{\alpha}_s$ $(\tilde{\alpha}_b)$ with probability one in equilibrium.*

Proof. Suppose that the buyer plays $\tilde{\alpha}_b$ with probability one in period t. If the probability that the seller plays a revealing offer in period $t + 1$ is nonzero, then the seller's strategy is suboptimal. Delaying this probability of playing a revealing offer until period $t + 1$ is strictly worse than allocating the probability of playing a revealing offer to period t, since, in the meantime, the buyer does not accept a concealing offer (cf. Lemma A5), make a revealing offer (by assumption), nor reveal information (since no randomization occurs). If an equilibrium is to exist, the seller must then play $\tilde{\alpha}_s$ with probability one in period $t + 1$. A similar analysis now reveals that the buyer must play $\tilde{\alpha}_b$ with probability one in period $t + 2$; the seller must play $\tilde{\alpha}_s$ with probability one in period $t + 3$; and so on. The game will then continue indefinitely, which is suboptimal. The case of the seller is similar.

A proof analogous to that of Lemma A5 is now given.

Lemma A7. *The soft seller (buyer) will play a revealing offer in period t with probability one iff $\pi_b^t \leq {}_{-1}\pi_b$ $(\pi_s^t \leq {}_{-1}\pi_s)$.*

Lemma A8. *Bargaining in the two-sided incomplete information game always concludes in a finite number of stages.*

Proof. In each period (with the possible exception of the first), the soft agent must be indifferent between playing a concealing and revealing offer. It is clear that if the soft seller's indifference is to be maintained, the probability that the soft buyer plays a revealing offer must not approach zero. However, this ensures that π_b^t will approach zero. In a finite number of periods, π_b^t will then fall below the strictly positive threshold $_{-1}\pi_b$. At this point (presuming π_s^t has not previously fallen below $_{-1}\pi_s$), the game will be ended by the soft seller's making a revealing offer.

Lemma A9. *A sequential equilibrium exists in the two-sided incomplete information game.*

Proof. Consider a restricted game. The game will be limited to a number of periods which is finite but longer than the finite number of periods examined in

Lemma A8. Agents are restricted to a finite set of offers given by $\tilde{\alpha}_s$ plus $\{x : x = x_s(\pi_b)$ for some $\pi_b \in [0, 1]\}$ for the seller, with a similar specification for a buyer (recall that $x_s(\pi_b)$ is a step function that takes on a finite number of values). This finite game will possess a sequential equilibrium (Kreps and Wilson). Next, we have seen that best-reply offers in the unrestricted game must come from the finite offer sets just described, and that the unrestricted game will not proceed beyond the finite number of periods of the restricted game. Hence, a collection of sequential best-reply strategies in the restricted game will also be a sequential equilibrium in the unrestricted game. This establishes existence in the unrestricted game.

Notes

1. The alternating-offers structure, popularized by Rubinstein (1982), has become a common formulation. However, the exogenously imposed timing of offers can be criticized. The obvious alternative is to construct a continuous-time model with no restrictions on the timing of offers. Unfortunately, such models generally allow so many equilibria as to be devoid of predictive power (cf. Stahl 1987).

2. Bargaining games with two-sided incomplete information and at most two stages have been studied by Binmore (1981), Chatterjee and Samuelson (1983), and Fudenberg and Tirole (1983). Infinite horizon models with one-sided incomplete information are examined by Bikhchandani (1985), Fudenberg, Levine, and Tirole (1985), Grossman and Perry (1986b), and Rubinstein (1985).

3. Optimistic conjectures cause our analysis of the one-sided incomplete information game to resemble that of Fudenberg, Levine, and Tirole, who impose a restriction that only the uninformed player make offers. Analyses that differ significantly from ours are offered by Bikhchandani and by Rubinstein (1985). The differences arise because the uncertainty in these games involves agents' discount factors, with a pie of known size to be divided.

4. To ensure that the revelation of an agent's type converts a game of two-sided (one-sided) incomplete information into a game of one-sided incomplete (complete) information, we assume that if an agent's belief that an opponent is soft becomes a probability of zero or one, then that belief is not subsequently revised. This causes a unitary probability of a soft opponent, for example, to be equivalent to common knowledge that the opponent is soft. This assumption is not a consequence of the sequential equilibrium

concept, though we find it to be an appealing condition and it is commonly invoked. For example, Grossman and Perry (1986a) include this condition as part of their perfect sequential equilibrium concept, and it also appears in Rubinstein (1985).

5. If the period t equilibrium offer is x_{nb}, any buyer offer $x > x_{nb}$ is accepted with probability $k_s(x, \pi'_s)$ such that if $x \in [x_{n'b}, x_{n'-2b})$ then $k_s(x, \pi'_s)$ causes π'_s to fall to $_{n'-2}\pi_s$. The sequence $_n\pi_s$ is again constructed in order to ensure such out-of-equilibrium offers are not profitable.

6. The necessity of the genericity qualification is now apparent. An equilibrium in which the seller randomizes first might exist even if $\pi^1_s < \zeta_s(\pi^1_b)$ if this randomization produces a point (π^2_s, π^1_s) on one of the intervals J_i, which are the counterparts of I_i for the case of the buyer. The set of initial conditions for which this is possible is clearly of measure zero.

7. The sequence of offers shown in Figure 2 contrasts with the equilibrium of the analogous complete information game examined by Rubinstein (1982). This is consistent with a growing collection of results which demonstrate that introducing seemingly small amounts of uncertainty can yield significant changes in equilibria.

Acknowledgment

We thank three anonymous referees and an associate editor for helpful comments and suggestions. Financial support from the National Science Foundation (SES-8419931) is gratefully acknowledged.

References

BIKHCHANDANI, S. 1985. A Bargaining Model with Incomplete Information. Mimeo, University of California, Los Angeles.

BINMORE, K. G. 1981. Nash Bargaining and Incomplete Information. Economic Theory Paper, University of Cambridge.

CHATTERJEE, K., AND W. F. SAMUELSON. 1983. Bargaining under Incomplete Information. *Opns. Res.* **31**, 835–851.

CHATTERJEE, K., AND L. SAMUELSON. 1987. Bargaining with Two-Sided Incomplete Information: An Infinite Horizon Model with Alternating Offers. *Rev. Econ. Studies* **54**, 175–192.

CHO, I.-K., AND D. M. KREPS. 1987. Signaling Games and Stable Equilibria. *Quart. J. Econ.* **102**, 179–222.

CRAMTON, P. C. 1984. Bargaining with Incomplete Information: An Infinite Horizon Model with Continuous Uncertainty. *Rev. Econ. Studies* **51**, 579–594.

CRAMTON, P. C. 1986. Strategic Delay in Bargaining with Two-Sided Uncertainty. Yale School of Organization and Management Working Paper No. 20 (Series D).

FUDENBERG, D., AND J. TIROLE. 1983. Sequential Bargaining under Incomplete Information. *Rev. Econ. Studies* **50**, 221–248.

FUDENBERG, D., D. LEVINE AND J. TIROLE. 1985. Infinite Horizon Models of Bargaining with Incomplete Information. In *Game Theoretic Models of Bargaining*, pp. 73–98, A. E. Roth (ed.). Cambridge University Press.

GROSSMAN, S., AND M. PERRY. 1986a. Perfect Sequential Equilibria. *J. Econ. Theory* **39**, 97–119.

GROSSMAN, S., AND M. PERRY. 1986b. Sequential Bargaining under Asymmetric Information. *J. Econ. Theory* **39**, 120–154.

KREPS, D. M., AND R. WILSON. 1982. Sequential Equilibria. *Econometrica* **50**, 863–894.

PERRY, M. 1986. An Example of Price Formation in Bilateral Situations: A Bargaining Model with Incomplete Information. *Econometrica* **54**, 314–322.

RUBINSTEIN, A. 1982. Perfect Equilibrium in a Bargaining Model. *Econometrica* **50**, 97–109.

RUBINSTEIN, A. 1985. A Bargaining Model under Incomplete Information. *Econometrica* **53**, 1151–1172.

STAHL, D. O., II, 1987. Bargaining in Continuous Time with Durable Offers. Mimeo, Duke University.